Cytotoxicity - New Insights into Toxic Assessment

Edited by Sonia Soloneski and Marcelo L. Larramendy

Published in London, United Kingdom

IntechOpen

Supporting open minds since 2005

Cytotoxicity - New Insights into Toxic Assessment
http://dx.doi.org/10.5772/intechopen.91554
Edited by Sonia Soloneski and Marcelo L. Larramendy

Contributors
Bhavin R. Chavda, Bhavesh N. Socha, Pavithra Sundaramoorthy, Kannan Kilavan Packiam, Harshavardhini Kuppusamy R, Jananipriya Srinivasan, Salud Pérez Gutiérrez, Nimsi Campos Xolalpa, Julia Pérez Ramos, Ana Esquivel Campos, Cuauhtemoc Pérez-González, Leonor Sánchez Pérez, Grethel León-Mejia, Alvaro Miranda-Guevara, Ornella Fiorillo Moreno, Carolina Uribe-Cruz, Elena Mikhailovna Mikhailovna. Egorova, Said Ibragimovich Kaba, Sonia Soloneski, Marcelo L. Larramendy

First published in London, United Kingdom, 2021 by IntechOpen
IntechOpen is the global imprint of INTECHOPEN LIMITED, registered in England and Wales, registration number: 11086078, 5 Princes Gate Court, London, SW7 2QJ, United Kingdom
Printed in Croatia

British Library Cataloguing-in-Publication Data
A catalogue record for this book is available from the British Library

Additional hard and PDF copies can be obtained from orders@intechopen.com

Cytotoxicity - New Insights into Toxic Assessment
Edited by Sonia Soloneski and Marcelo L. Larramendy
p. cm.
Print ISBN 978-1-83968-726-6
Online ISBN 978-1-83968-727-3
eBook (PDF) ISBN 978-1-83968-728-0

For EU product safety concerns:
IN TECH d.o.o., Prolaz Marije Krucifikse Kozulić 3, 51000 Rijeka, Croatia, info@intechopen.com or visit our website at intechopen.com.

Meet the editors

Sonia Soloneski has a Ph.D. in Natural Sciences and is Assistant Professor of Molecular Cell Biology at the School of Natural Sciences and Museum of La Plata, National University of La Plata, Argentina. She is a member of the National Scientific and Technological Research Council (CONICET) of Argentina in the Genetic Toxicology field, the Latin American Association of Environmental Mutagenesis, Teratogenesis and Carcinogenesis (ALAMCTA), the Argentinean Society of Toxicology (ATA), the Argentinean Society of Biology (SAB) and the Society of Environmental Toxicology and Chemistry (SETAC). She has authored more than 380 contributions in the field, including scientific publications in peer-reviewed journals and research communications. She has served as a review member for more than 30 scientific international journals. She has been a plenary speaker in scientific conferences and a member of scientific committees. She is a specialist in issues related to Genetic Toxicology, Mutagenesis, and Ecotoxicology.

Marcelo L. Larramendy, Ph.D., serves as Professor of Molecular Cell Biology at the School of Natural Sciences and Museum (National University of La Plata, Argentina). Appointed Senior Researcher of the National Scientific and Technological Research Council of Argentina. Former Member of the Executive Committee of the Latin American Association of Environmental Mutagenesis, Teratogenesis and Carcinogenesis. Author of more than 450 contributions, including scientific publications, research communications and conferences worldwide. Recipient of several national and international awards. Prof. Larramendy is a regular Lecturer at the international A. Hollaender Courses organized by the IAEMS and former guest scientist at NIH (USA) and the University of Helsinki, (Finland). He is an expert in Genetic Toxicology and is, or has been, referee for more than 20 international scientific journals. Member of the International Panel of Experts at the International Agency for Research on Cancer (IARC, WHO, Lyon, France) in 2015 for the evaluation of DDT, 2,4-D and Lindane. Presently, Prof. Dr. Larramendy is Head of the Laboratory of Molecular Cytogenetics and Genotoxicology at the UNLP.

Contents

Preface

Despite advances in science, a huge number of untested substances remain, necessitating that priorities are established for evaluating toxic inducers prior to their approval for animal and human uses. The toxicity of many xenobiotics is well-known today and interest in understanding the action mechanisms is being continuously stimulated by their continual re-evaluation. In this way, one of the major goals of cytotoxic estimation includes the preliminary screening of potential xenobiotics to determine the toxicity and efficacy of any substance, product or environmental agent, through employing *in vitro* approaches. These types of analyses are preferred in pilot testing in fields such as health, pharmaceuticals, environment, industry, agriculture and food, in order to estimate cell growth, reproduction and morphological effects.

Accurate adverse assessment by cytotoxic estimation as well as their role in different biological systems are a primary step employing for ranking the safety of many chemicals. In this regard, knowledge of relative toxicity is essential in order to determine a chemical's fate and thereby prevent or minimize after-effects. Similarly, the identification of cytotoxic responses may be essential for elucidating target cells and organ toxicity.

This book is intended to present some strategies, methods, interpretations and recent advances in order to facilitate scientific research on *in vitro* toxic responses, presenting both theoretical and practical aspects.

Although vast literature is available on cytotoxic evaluation, this book contains important investigations into the diverse chemical hazards encountered in anthropogenic and natural environments. Moreover, it provides valuable information regarding the toxicity of several xenobiotics that can negatively affect human beings. The contributors clearly discuss several concepts and approaches that will be useful for understanding the potential action mechanisms of various compounds, namely quinones, silver nanoparticles, antibiotics and plant extracts, among others.

This book is organized into six chapters. It begins with an introductory chapter that presents an overview of different approaches recommended for the preliminary *in vitro* screening of cellular response and analyses of dead cells within a cell population. The chapter highlights the strengths and weaknesses of some cytotoxic endpoints routinely employed and shows how the selection of an appropriate cytotoxicity bioassay is decisive in order to obtain a comprehensive toxicity profile. The second chapter presents a complete study highlighting the different responses of a cell to several xenobiotic agents as well as different tests that can be useful for evaluating cellular responses. The chapter focuses on the toxicological and cytotoxic evaluation of chemical substances through *in vitro* tests, as a competitive alternative to *in vivo* experimentation as a consequence of ethical considerations. The third chapter reviews the role of quinones – an important family of natural products mainly isolated from bacteria, fungi, algae, plants and other organisms – as possible agents to prevent cancer and microbial activity. The chapter summarizes the activities of 152 anticancer and 30 antimicrobial quinones, showing that these categories of compounds exert an excellent response against numerous cell cancer lines and also exhibit good antimicrobial activity, determined by *in vitro* and *in vivo* studies. The fourth chapter discusses the impact of some methodological strategies in toxicity studies of silver nanoparticles on cultured cells. The chapter highlights the employment of detailed conditions required to correctly determine the nanoparticle size effect in studies of living cells and the invalidity problem of Smoluchowski's equation. It then describes an approach to analyze cytotoxicity when nanoparticles are stabilized with various surfactants. The fifth chapter provides toxic information on the assessment by lethal concentration 50 of a bacteriostatic antibiotic, namely sulfamethoxazole and its silver-sulfamethoxazole complex, employing brine shrimp as an experimental model. The final chapter discusses the medicinal properties of the plant *Evolvuluous alsinoides*, including multiple potential activities like antibiotic, antimicrobial, anti-inflammatory and antidiabetic properties. Furthermore, the chapter includes a brief compilation of scientific findings showing the protective role of phytochemicals under several pathological conditions.

The editors of ***Cytotoxicity - New Insights into Toxic Assessment*** are enormously grateful to all the contributing scientists for sharing their knowledge and insights in this interdisciplinary book project. They have made an extensive effort to arrange the information included in every chapter. The publication of this book is of high importance for students, researchers and scientists in diverse fields with expertise in toxicology, health, pharmaceuticals and other disciplines, who can contribute and share their findings to take this area forward for future investigations.

Sonia Soloneski Ph.D. and Marcelo L. Larramendy Ph.D.
School of Natural Sciences and Museum,
National University of La Plata,
La Plata, Argentina

Chapter 1

Introductory Chapter: Cytotoxicity

Sonia Soloneski and Marcelo L. Larramendy

1. Introduction

The evaluation of toxicity is an important process for assessing the hazards and risks that diverse xenobiotic released into the environment have for human beings, animals, plants and all compartments in our environment. In particular, the toxic response at cellular level plays a central role in the identification and quantification of adverse outcomes associated with exposure to numerous pollutants, including natural toxins, food additives, pesticides, nanomaterials, metals, radiation, viruses, biomolecules and medical devices, among others [1–5].

Conventional *in vivo* testing frequently requires a huge number of animal experiments. However, after application of Russel and Burch's strategy based on the '3Rs' – reduction, refinement and replacement – the replacement of animals of experimentation for research purposes has been considered, to minimize the cruelty and misuse that provoke pain, distress and death experiences [6]. Strategies employing specialized animal/human cell lines, tissue cultures, callus cultures, organ cultures and the continued use of long-established cultures are an excellent and practical way to screen the properties of any xenobiotic in the early stages of experimentation. It is well known that *in vitro* cytotoxicity cannot replace the conventional *in vivo* effects detected in the advanced stages of product development but cytotoxic estimation will assist in extrapolating *in vitro* observations to predict or at least to suggest a clue of the *in vivo* effects [7].

Accurate assessment of the adverse effects of xenobiotics by estimating their cytotoxicity as well as their role in different biological systems is a primary step employed to rank the safety of many chemicals; knowledge of the relative toxicity is essential in order to decide the fate of a chemical to prevent or minimize their effects and to identify cytotoxic responses that may be essential for elucidating target cells as well as organ toxicity. According to Freshney [8], the meaning of cytotoxicity can differ, considering upon the nature of the study and whether cells are killed or simply have their metabolism altered. For example, whereas an anticancer drug employed in chemotherapy may be required to kill cells, the absence of toxicity in other chemicals may involve a deep complete analysis of specific targets such as modifications in cell signaling or cell interaction, among other deleterious effects [8].

In recent years, science and technology innovations have accelerated the progress in the standardization of methods for determining cytotoxicity that are properly sensitive to predict several levels of cell toxicity, i.e., from low to high. These bioassays are efficient and economical tools that can quickly make valuable responses that are suitable for both qualitative and quantitative assessment [8].

There are numerous as well as highly recommended methodologies routinely used for preliminary *in vitro* screening of cellular response and the analysis of dead cells within a cell population. Advantages associated with *in vitro* approaches are that they are easy to follow, less time-consuming and less expensive than other developmental

 IntechOpen

designs [9]. The selection of an appropriate cytotoxicity bioassay is decisive for estimating xenobiotic toxicity and it may be necessary to perform several methodologies at the same time to obtain a comprehensive toxicity profile due to numerous sublethal cellular changes that may occur after a short period of exposure [10].

Cytotoxic evaluation employs several endpoints such as cell viability, cell cycle function and control, cell membrane integrity, DNA synthesis, metabolic side effects and apoptosis, among others, as indicators that can potentially help to determine cellular damage and viability. There are a lot of different methods for estimating cytotoxicity. Each approach comes with its own set of strengths and weaknesses regarding its specificity and sensitivity. Therefore, depending on the study and targeted endpoint, an appropriate bioassay should be selected. So far, there is no single method alone which has been found to be a suitable indicator of cytotoxicity. Several methodologies such as the neutral red uptake (NRU) assay for estimating cell viability and membrane damage, the Coomassie blue and Kenacid blue assays for measuring cellular proliferation and total protein content, tetrazolium-based colorimetric assays as indicators of mitochondrial function, and the cellular leakage of lactate dehydrogenase for measuring cell injury are the most commonly and habitually employed worldwide [8].

2. Cell viability assays

Numerous cell viability and cell proliferation bioassays are routinely employed to analyze the toxicity profile of a xenobiotic on cells cultured *in vitro*. Estimation of the proportion of viable cells relies on an interruption of cellular membrane integrity by the incorporation of a dye after a chemical treatment, i.e., dye exclusion and preferential dye uptake are frequently employed to distinguish and quantify the proportion of live cells in suspension. Numerous vital dyes, including propidium iodide, trypan blue, methylene blue, erythrosine B, nigrosine, eosin, safranin, naphthalene black, 7-aminoactinomycin D, and Hoechst 33342 (bis-benzimide H 33342 trihydrochloride) have been introduced to estimate the proportion of viable cells [8].

The NRU is the most performed colorimetric bioassay employed to estimate cell viability and cytotoxicity. Living cells take up and bind the supravital dye 3-amino-7-dimethylamino-2-methylphenazine hydrochloride – neutral red – and sequester it in the lysosomes and endosomes. The uptake of neutral red depends on the cell's ability to preserve pH gradients, through the production of ATP, and neutral red is not retained by dead cells. This bioassay does not estimate the total number of cells, but it detects only viable cells [11–13].

The Coomassie blue assay is a methodology employed for determining total protein content based on differential binding of the stain by the protein and the matrix under acidic conditions. Analysis of cellular proliferation and total protein content based on Coomassie blue staining represents a quick, simple, and affordable method for detecting cytotoxicity [14]. The cytotoxic effect of xenobiotics can be also estimated by modifications in total cell protein by the Kenacid blue dye binding assay. The basis of this method is that a test chemical will interfere with this process and thus result in a reduction of the growth rate as reflected by cell number. The degree of growth inhibition, related to the concentration of the test compound, provides an indication of toxicity [15].

The lactate dehydrogenase (LDH) bioassay is another colorimetric methodology employed for assaying cellular toxicity. LDH is a cytosolic enzyme present in many different cell types and is a well-defined and reliable indicator of cytotoxicity. Alterations in the cell membrane provoke the release of cytosolic contents, including LDH enzyme, into the cell culture medium. The amount of extracellular LDH can be estimated using a colorimetric assay in which the amount of product formed, a tetrazolium salt reduced to a red formazan, correlates to the amount of dead or damaged cells [16].

Bioassays that estimate the proportion of viable cells indirectly, by analyzing the reduction of the intracellular environment and employing metabolic biomarkers, are suitable and offer fewer disadvantages than other available methodologies. However, one possible disadvantage of some of these bioassays is that there is no distinction between cells that are in proliferation and those that are quiescent, which may result in overestimation of the number of analyzed cells. The most representative is the metabolic bioassay for estimating ATP (adenosine triphosphate), indicative of cell survival and cell growth and determining morphology. ATP is a ubiquitous carrier of chemical energy and bioassays that quantify intracellular ATP levels indicate cell death [17]. Other metabolic bioassays determine NAD^+ (nicotinamide adenine dinucleotide) and $NADP^+$ (nicotinamide adenine dinucleotide phosphate), two ubiquitous soluble cofactors which are found in cells and implicated in energy metabolism, signal transduction and cellular homeostasis. In cells, both cofactors are present as oxidized and reduced dinucleotide forms, and changes in redox environment are employed to analyze the proportion of viable cells [18].

Metabolic tetrazolium-based colorimetric assays such as the MTT (3-(4,5-dimethylthiazol-2-yl)-2–5-diphenyltetrazolium bromide), MTS (5-(3-carboxymethoxyphenyl)-2-(4,5-dimethyl-thiazoly)-3-(4-sulfophenyl) tetrazolium, inner salt assay) and XTT (2,3-bis(2-methoxy-4-nitro-5-sulphophenyl)-5-carboxanilide-2H-tetrazolium, monosodium salt) bioassays are designed to estimate cytotoxicity by measuring the reduction of a colorimetric substrate associated not only with mitochondria, by mitochondrial enzymes such as succinate dehydrogenase, but also with the cytoplasm and with non-mitochondrial membranes including the endosome/lysosome compartment and the plasma membrane. The MTT, MTS and XTT bioassays quantify the proportion of viable cells using a colorimetric assay in which the bioreduction of a tetrazolium salt to an intensely colored formazan correlates to the amount of dead or damaged cells determined by measuring absorbance at 450 nm [19].

3. Perspectives

Cytotoxicity biomarkers have proved most useful as tools to elucidate the biochemical and/or metabolic changes involved in the toxic action mechanisms of xenobiotics at the cellular level. Many approaches have been optimized and refined, more multicentre protocols have been performed and international analysis and exchange of information have considerably increased. The increasing collection and evaluation of cytotoxic biomarkers is also providing growing opportunities and numerous challenges for regulatory toxicity testing, motivating the employment of fewer animals of experimentation.

This book, *Cytotoxicity*, is intended to present the rationale, strategies, methods, interpretations and recent advances in *in vitro* toxicity, presenting both theoretical and practical aspects.

Acknowledgements

This study was supported by grants from the National University of La Plata (Grants 11/N847 and 11/N926) and the National Agency of Scientific and Technological Promotion (PICT 2015 Number 3059) from Argentina.

Conflict of interest

The authors declare that there are no conflicts of interest.

Author details

Sonia Soloneski[1,2]* and Marcelo L. Larramendy[1,2]

1 Cátedra de Citología, Facultad de Ciencias Naturales y Museo, Universidad Nacional de La Plata, La Plata, Argentina

2 Consejo Nacional de Investigaciones Científicas y Técnicas (CONICET), Argentina

*Address all correspondence to: ssoloneski@yahoo.com.ar

References

[1] FAO. Pesticide Residues in Food - Evaluations Part 1: Residues. FAO Plant Production and Protection 152/1, 1988; p. 179-189.

[2] Brunner T J, Wick P, Manser P, Spohn P, Grass R N, Limbach L K, Bruinink A, Stark W J. *In vitro* cytotoxicity of oxide nanoparticles: comparison to asbestos, silica, and the effect of particle solubility. Environ. Sci. Technol. 2006; 40: 4374-4381.

[3] Chatterjee S, Sarkar S, Bhattacharya S. Toxic metals and autophagy. Chem. Res. Toxicol. 2014; 27:1887-900.

[4] Liu X, Rodeheaver D P, White J C, Wright A M, Walker L M, Zhang F, Shannon S. A comparison of *in vitro* cytotoxicity assays in medical device regulatory studies. Regul. Toxicol. Pharmacol. 2018; 97:2 4-32.

[5] Patravale V, Dandekar P, Jain R. Nanotoxicology: Evaluating Toxicity Potential of Drug-Nanoparticles, In: Patravale, V, Dandekar, P, Jain, R, editors. Nanoparticulate Drug Delivery: Perspectives on the Transition from Laboratory to Market Woodhead Publishing Limited, Cambridge, United Kingdom, 2012; p. 123-155.

[6] Russell W M S, Burch R L, editors. The principles of humane experimental technique. London: Methuen & Co. Limited, 1959; p. 252.

[7] Das S. Extrapolation of *In Vitro* Results to Predict Human Toxicity. In: Dhawan A, Kwon S, editors. *In Vitro* Toxicology. Academic Press, 2018; p. 127-142.

[8] Freshney R I. Cytotoxicity. In: Culture of Animal Cells: A Manual of Basic Techniques, 5th ed. John Wiley & Sons, New York, 2005; p. 359-376.

[9] Doke S, Dhawale S. Alternatives to animal testing: A review. Saudi Pharm. J. 2015; 23: 223-229.

[10] Aslantürk, Ö S. *In Vitro* Cytotoxicity and Cell Viability Assays: Principles, Advantages, and Disadvantages. In: Larramendy M L, Soloneski S, editors. Genotoxicity - A Predictable Risk to Our Actual World. IntechOpen, 2017; p. 1-6

[11] Ates G, Vanhaecke T, Rogiers V, Rodrigues, R M. Assaying Cellular Viability Using the Neutral Red Uptake Assay, In: Gilbert, D F, editors. Cell Viability Assays. Methods Mol. Biol. Human Press, Springer, New York, New York, 2017; p. 19-26.

[12] Borenfreund E, Puerner J A. Toxicity determined *in vitro* by morphological alterations and neutral red absorption. Toxicol. Lett. 1985; 24, 119-124.

[13] Borenfreund E, Babich H, Martin-Alguacil, N. Comparisons of two *in vitro* cytotoxicity assays-The neutral red (NR) and tetrazolium MTT tests. Toxicol. *In Vitro*. 1988; 2:1-6.

[14] Brunelle J L, Green R. Coomassie blue staining. Methods Enzymol. 2014; 541:161-167.

[15] Clothier R H. The FRAME cytotoxicity test (Kenacid Blue). In: Hare S O and Atterwill C, editors. *In Vitro* Toxicity Testing Protocols. Methods in Molecular Biology. Chapman and Hall, London, United Kingdom, 1995; p. 109-118.

[16] Decker T, Lohmann-Matthes M. A quick and simple method for the quantitation of lactate dehydrogenase release in measurements of cellular cytotoxicity and tumor necrosis factor (TNF) activity. J. Immunol. Meth. 1988; 15: 61-69.

[17] Eguchi Y, Shimizu S, Tsujimoto Y. Intracellular ATP levels determine cell death fate by apoptosis or necrosis. Cancer Res. 1997; 57: 1835-1840.

[18] Ying W. NAD/NADH and NADP/NADPH in cellular functions and cell death: Regulation and biological consequences. Antioxid. Redox Signal. 2007; 10: 179-206.

[19] Berridge M V, Herst P M, Tan A S. Tetrazolium dyes as tools in cell biology: new insights into their cellular reduction. Biotechnol. Annu. Rev. 2005; 11:127-152.

Chapter 2

Cytotoxicity as a Fundamental Response to Xenobiotics

Grethel León-Mejía, Alvaro Miranda Guevara, Ornella Fiorillo Moreno and Carolina Uribe Cruz

Abstract

Cytotoxicity refers to the ability of a molecule or a compound to cause some type of cellular damage, of which some of the adverse effects that can occur include injuries to some structures or the fundamental processes involved in cell maintenance, such as survival, cell division, cell biochemistry, and the normal cell physiology. The potential for cytotoxicity is one of the first tests that must be performed to determine the effects of drugs, biomolecules, nanomaterials, medical devices, pesticides, heavy metals, and solvents, among others. This potential may be oriented in the mechanism under which it generates cell death, the dose, and the target cells that generate the response. The evaluation of the toxicologic and cytotoxic properties of the chemical substances through *in vitro* tests has become a competitive alternative to *in vivo* experimentation as a consequence of ethical considerations. Presently, there are numerous tests conducted to evaluate the cytotoxicity of a certain agent, the selection of which depends on the purpose of the study. In this sense, the present review provides a general overview of the different responses of a cell to xenobiotic agents and the different test that can be useful for evaluation of these responses.

Keywords: cytotoxicity, xenobiotic, apoptosis, biomarkers, cellular damage

1. Introduction

In the modern world, increasing industrialization continues to pose serious pollution problems [1]. Every year, several countries generate millions of tons of pollutants, which keeps adding to this interrelation among population, technology, resource consumption, and the environment—a situation which is becoming increasingly complex [2]. Although it is true that the reality is alarming, risk quantification and estimation strategies, based on studies of the effects of xenobiotics, are necessary not only for the conservation of the environment but also to acquire the knowledge about the factors that intervene in each specific case to enable foreseeing potential injuries [1]. Humans are exposed to a wide variety of foreign chemical substances, which we collectively call "xenobiotics" that includes natural compounds present in plant foods, such as synthetic compounds in medicines, food additives, and environmental pollutants. At present, the study on the effects of xenobiotics and the elucidation of their mechanisms of action on macromolecules has become extremely important [3, 4].

Cytotoxicity is defined as an alteration of the basic cellular functions that leads to detectable damage [5]. In this sense, the cytotoxicity that a xenobiotic induces may be a relevant point in the elucidation of the mechanism of action of the xenobiotic [6]. The analysis of cytotoxic effects is a fundamental strategy involved in the analysis of the xenobiotic–cell interaction in basic research, industrial development, and in the evaluation of therapeutic and toxic effects of chemical and biological products [5].

2. Cytotoxicity and genotoxicity

Genotoxicity is defined as the capacity of an agent, be it physical, chemical, or biological, to cause damage to the genetic materials or to alter the cellular components that influences the functionality and behavior of the chromosomes within the cell, leading to adverse biological outcomes [7]. Hence, the final cause of cell death may be related to DNA damage [8]. Specific cellular responses include cell cycle arrest and attempts to repair DNA [9]. The use of a specific repair enzyme complex depends on the type of DNA strand break or the chemistry of the adduct formed as well as the repair capacity of the affected cells [9]. If there is excessive DNA damage, an alternative is the action of p53 that activate apoptosis [8].

Apoptosis

One of the most obvious end points of action for several drugs and toxic xenobiotics is cell death [10]. Cell death is divided into two types: i) necrosis, which is characterized by the occurrence under the mechanisms of irreversible cell injury and is considered accidental and ii) apoptosis, which corresponds to programmed cell death that runs under control and is related to homeostasis of tissue growth [11]. The term programmed cell death was introduced for the first time in 1920, and, in 1972, the term apoptosis was coined by Wyllie and Currie, since then the mechanisms and molecular aspects by which this process is conducted have been described [12].

The extrinsic signaling pathway that initiates apoptosis has been so named since it involves interactions mediated by transmembrane receptors, and it has been described that this pathway is initiated by the binding of i) tumor necrosis factor (TNF) ligand to the receptor TNF, ii) TNF-related apoptosis inducing ligand (TRAIL) to death receptor-4 (DR4), and DR5 receptors, or 3) the fatty acid synthetase ligand to the FasR receptor. These associations are known to recruit adapter molecules such as Fas-associated death domain (FADD) or TNF receptor-associated death domain (TRADD), which activates initiator caspases-8 and -10 and, finally, the activation of executor caspases-3, −6, and − 7 that culminates in an apoptotic cell phenotype with characteristic physiological and morphological characteristics [11, 13].

In the intrinsic signaling cascade, a series of intracellular stimuli has been reported specifically in the mitochondria, which induces structural changes in the mitochondrial membrane, mainly due to the opening of the transition pores and alterations in the transmembrane potential entailing a release toward the cytosol of pro-apoptotic substances, which remains within the intermembranal space in a normal state [14]. These released components have been classified into two main groups consisting i) cytochrome c, Smac/diablo (second mitochondrial activator of caspases/direct inhibitors of apoptotic proteins (IAP) binding protein with low Propidium iodide (PI)) and the serine protease HtrA2/Omi (high-temperature requirement): these releases lead the apoptotic cascade via caspase activation. It has been discovered that the release of cytochrome c activates the apoptotic protease activating factor-1 (Apaf-1 protein) and procaspase-9 in addition to ATP, thereby establishing a protein complex called as the apoptosome, which in turn activates caspase-3, initiating the effector pathway of apoptosis. On the other hand, the Smac/Diablo and HtrA2/Om proteins promote apoptosis by inhibiting IAP such as cIAP1, cIAP2, and XIAP. ii) Apoptosis inducing factors, endonuclease G, and caspase-activated DNase: the latter are released into the cytosol, enter the nucleus, and fragment the DNA. The importance of DNA degradation by Ca^{2+} and Mg^{2+} dependent endonucleases is that they generate fragments of 180–200 base pairs; this pattern of fragments is highly specific and an extremely clear factor that differentiates this type of programmed cell death with necrosis, which does not present a degradation pattern or specificity in the fragment sizes [11, 13, 14].

Autophagy

Autophagy is an intracellular degradation process that is characterized by the formation of double membrane vesicles called as autophagosomes; these vesicles sequester the cytoplasmic material and later fuse with the lysosome (that contains hydrolytic enzymes), forming the autophagolysosome or autolysosome, where the degradation of the invaginated material occurs [15]. The amino acids and small molecules that are generated through autophagy are returned to the cytoplasm for the generation of energy and for the synthesis of new proteins and biomolecules [16]. The main inducer of autophagy is nutrient deficiency; however, it has been reported that the activation of autophagy is a cell survival mechanism against various stress conditions, including oxidative stress, inflammation, protein aggregation, endoplasmic reticulum stress, metabolic stress, the presence of pathogens, and changes in the mitochondrial function [17]. Autophagy plays an important role in cell and tissue homeostasis by contributing to the generation of energy from degradation events, which plays a role in the quality control of proteins and organelles, in the elimination of long-lived proteins and pathogens, as well as in the regulation of cell death [15, 16].

The central machinery of the autophagy process is composed of >30 proteins, including the so-called Atg. The autophagy pathway proceeds through five phases: (i) nucleation, which is the formation of a double membrane structure or an isolating membrane called the "phagophore"; (ii) the expansion of the phagophore membrane by the incorporation of the LC3-II protein; (iii) the maturation of this structure in the autophagosome and the sequestration of cytoplasmic material to be degraded; (iv) the fusion of the autophagosome with the lysosomes, which results in the formation of autophagosome and autolysosomes and, finally; (v) the degradation of biological materials sequestered by the hydrolytic enzymes of the lysosome and the recycling of the molecules (above all amino acids, lipids, sugars, and nucleotides) [17, 18].

The nucleation and formation of the phagophore is initiated by the serine and threonine kinase unc-51-like autophagy activating kinase 1 (ULK1). Once the ULK1 complex is activated, it activates the phosphatidylinositol 3-kinase (PtdIns3K) complex that includes members such as Beclin 1, Atg14, Vps15, Vps34, and Ambra1. Two ubiquitin-like conjugation systems participate in the expansion of the phagophore membrane, until the closure of the double membrane vesicle to form the mature autophagosome, which are Atg12-Atg5-Atg16 and Atg4-Atg7-Atg3/LC3-PE (phosphatidylethanolamine) [17, 18]. Eventually, the autophagosome fuses with the lysosome, and the sequestered material is degraded by lysosomal enzymes (such as cathepsins, glucosidases, lipases, and sulphatases). The components of degraded biomolecules, for example amino acids, are returned to the cytoplasm to derive energy and for the synthesis of new biomolecules [17, 18].

3. Cytotoxicity test

Within the battery of *in vitro* tests that are useful and necessary, alternative toxicology methods for the registration or application of clinical trials of a given substance are referred to as cytotoxicity tests; these tests are capable of detecting, through different known cellular mechanisms, the adverse effects of interference with structure and/or properties essential for cell survival, proliferation, and/or functions [19, 20]. These tests include the integrity of the membrane and the cytoskeleton, metabolism, synthesis, degradation, release of cellular constituents or products, ionic regulation, and cell division [19, 20].

The sensitivity and speed of the damage analysis at the cellular level increases its practical value when simple cytotoxicity markers are used, such as in the determination of viability by exclusion of fluorescent and non-fluorescent dyes. There is a wide availability of markers for intracellular and extracellular structures and functions, as well as to examine several of these markers simultaneously in the same cell to allow analysis at individual cell level [19, 20].

Among the best-known tests are neutral red uptake assay, 3-(4,5-dimethylthiazol-2-yl)-2,5-diphenyltetrazolium Bromide (MTT) reduction, clonogenic assay, sulforhodamine B, lactate dehydrogenase (LDH) release colorimetric assay, annexin V/propidium iodide (PI) staining, TUNEL, and kenacid blue and resazurin binding assay. **Table 1** shows *in vitro* key studies on cytotoxicity induced by xenobiotics using different assay.

Neutral red uptake assay

This test is a measure of the toxicity of a compound in the short or long term, which is determined by the release of a dye (i.e., neutral red) due to the loss of cell viability. In this sense, the fact that a compound is cytotoxic regardless of its action mechanism must be considered, if this interferes in the process of cell division and multiplication [34, 35]. This interference leads to the reduction in the speed of cell growth, which is reflected in the number of cells present in the culture. The degree of growth inhibition related to the concentration of the compound being evaluated is an index of toxicity [34, 35].

Neutral red is taken up by cells (specifically by lysosomes and endosomes) and, as the cell loses viability due to the action of the compound being evaluated, the dye is released into the medium, since only viable cells can retain the dye inside. The amount of neutral red dye that remains after exposure within the cell is then determined and the concentration that produces 50% inhibition of cell growth is then calculated [34, 35].

The MTT reduction assay

This method is simple and is used for the determination of cell viability, given by the number of cells present in the culture, which can be measured based on the formation of a colored compound as a result of a reaction occurring in the mitochondria of the viable cells [35, 36].

MTT is a compound belonging to the family of tetrazolium salts that is soluble in water and a yellow color. The metabolic activity of cells includes mitochondrial succinic dehydrogenase enzyme, but cytosolic reductases or reductases from other subcellular compartments may also be involved. The resulting reduced coenzymes (NADH and NADPH) will convert MTT to its insoluble formazan form [37]. When reduced, MTT becomes a compound purple and insoluble in water. To quantify MTT, it is usually dissolved in an organic solvent such as dimethyl sulfoxide (DMSO). The amount of reduced MTT is quantified by a colorimetric method, since the color changes from yellow to purple as a result of the reaction [35, 37, 38].

Reference	Xenobiotic	Cells used	Assay used	Outcome(s)
[21]	Pesticides Deltamethrin Fenitrothion, Fipronil, Lambda-cyalothrine, and Teflubenzuron	Caco-2 cells	MTT cell viability assay	Cytotoxic effect of Deltamethrin, Fenitrothion, Fipronil, Lambda-cyalothrine, and Teflubenzuron alone or in combination in human intestinal Caco-2 cells.
[22]	Malathion	N2a mouse neuroblastoma cells	MTT cell viability and LDH release assay	The non-cholinergic effect of malathion may be mediated by apoptotic cell death via autophagy and lysosomal membrane permeabilization induction in N2a cells.
[23]	X-ray	Human hepatocellular carcinoma cells	MTT and clonogenic assays	Cell autophagy was significantly increased after ionizing radiation combined with hyperthermia treatment. Autophagic cell death may be due to the increased intracellular ROS.
[24]	Gamma radiation	Human breast cancer cell line (MCF-7)	Clonogenic cell survival assay, cell viability using trypan blue staining and apoptotic cell death using the TUNEL assay	The dose and time dependence inducing a significant apoptotic death.
[25]	Heavy metals	Human sperm cells	WST-1 and XTT	Harmful effect of $CuSO_4$ and $CdCl_2$ on human spermatozoa.
[26]	Heavy metals	HT-22 cell line	MTT assay and Annexin V-FITC/ Propidium iodide (PI)	Metal mixtures showed higher cytotoxicity compared to individual metals.
[27]	Chemotherapeutic drug Cyclophosphamide	Monocyte Macrophage Cell Line Raw 264.7	MTT assay	A reduction in cell viability was found in Raw 264.7 cell line indicating the cell cytotoxicity.
[28]	Chemotherapeutic drugs Paclitaxel Docetaxel Oxaliplatin Bicalutamide Anastrozole	HT-29 and HeLa cells	MTT Assay	Dose-response cytotoxicity findings. Favorability of *in vitro* assay for the selection of chemotherapeutic drugs for greater clinical effectiveness.
[29]	Aminated polystyrene, zinc oxide, and silver nanoparticles	HeLa cells	MTT, Alamar blue, and neutral red assay	All nanoparticles tested resulted in the decrease in cell viability, increased intracellular ROS production and induction of cell death by caspase-mediated apoptosis.

Reference	Xenobiotic	Cells used	Assay used	Outcome(s)
[30]	Copper oxide, copper-iron oxide, and carbon nanoparticles	Human hepatoma HepG2 cells	MTT and neutral red assays	There was increased cytotoxicity, mutagenicity, and mitochondrial impairment in the cells treated with higher concentrations of the nanomaterials, especially the copper oxide nanoparticles.
[31]	Dental universal adhesives	Monocyte/ macrophage peripheral blood cell line	XTT assay	Some of the tested adhesives showed significant cytotoxic and genotoxic effects.
[32]	Perfluorocarbons for intraocular use	BALB/3T3, ARPE-19 cell lines	MTT, neutral red uptake, and TUNEL assay	Qualitative evaluation showed that cytotoxic control induced apoptosis, severe reactivity zones, and cytotoxicity according to ISO 10993-5 in all tested conditions.
[33]	Polysiloxane-based polyurethane/ lignin elastomers	HeLa cells	MTT assay	Demonstrate the usefulness of in vitro cytotoxicity studies to improve the response of materials based on polysiloxane-based polyurethane / lignin elastomers.

Table 1.
Overview of in vitro key studies on cytotoxicity induced by xenobiotics using different assay.

Clonogenic assay

This assay enables assessing whether a cell is capable of dividing and forming a colony after being exposed to a treatment [39]. A cell survival curve defines the relationship between the dose of the agent used and the fraction of cells that retain their ability to reproduce. Cell lines of various origins, neoplastic or normal, of human or rodent origin can be used [19, 39].

This test is considered an extremely useful tool owing to its advantages of low cost, reproducibility, and simplicity. It has been used for several decades to evaluate the effects of radiation, chemotherapy, drug development, drug screening, toxicology, and pharmacology [19, 40].

Sulforhodamine B

Sulforhodamine B (SRB) is a bright pink aminoxanthan dye with two negatively charged sulfonic groups $-SO_3-$ capable of electrostatically binding to cations [41]. Under acidic conditions (when dissolved in 1% acetic acid), SRB increases its affinity for the basic amino acids of proteins and binds selectively to them, providing an index of the cellular protein content if the cells were previously fixed with trichloroacetic acid. After removing the unfixed dye, the dye bound to the viable cells is extracted with alkaline medium (Tris solution, pH 10.5) and the absorbance is read at 564 nm [41, 42].

LDH release colorimetric assay

This assay allows the measurement of LDH enzyme activity using a cocktail of reagents containing lactate, NAD+, diaphrose, and the tetrazolium salt INT [43]. LDH catalyzes the reduction of NAD+ to NADH in the presence of L-lactate, and the formation of NADH can be measured by a coupled reaction, in which the tetrazolium salt INT is reduced to a red formazan product that can be measured spectrophotometrically [44]. The increase in the LDH activity in the culture supernatant is proportional to the number of cells lysed [43, 44].

Annexin V/PI staining

Annexin V is a recombinant protein that specifically binds to phosphatidylserine residues, which are exposed on the outer surface of the plasma membrane, and is an effective biomarker in apoptotic cells [45]. Annexin V can be combined with a DNA marker that is not membrane-permeable unless the membrane is compromised in order to distinguish apoptotic cells from necrotic cells [45].

It has been reported that the combination of annexin V-FITC and the cationic marker PI can guarantee this differentiation, registering non-apoptotic cells (annexin V-FICT-negative/PI negative), the cells in early apoptosis (annexin V-FICT positive/PI negative) and necrotic cells (annexin V-FICT positive/PI positive). The samples are analyzed in a cytometer providing an objective and fast quantification [46].

TUNEL assay

A useful method to study apoptosis is the TdT-mediated dUTP-biotin nick end-labeling (TUNEL assay). During apoptosis, nuclear endonucleases digest genomic DNA into oligonucleosomal fragments of approximately 180–200 base pairs. DNA fragments are labeled by the catalytic incorporation of labeled 16-dUTP at the free ends by means of the enzyme terminal deoxynucleotidyl transferase (TdT) [47]. The accessibility of the enzyme to the DNA break points is decreased due to nuclear proteins along with the processes of fixation of the sample and the subsequent fixation with ethanol [48]. The signal increases with a greater number of breaks in the DNA chain, and it can be conducted both by flow cytometry and fluorescence. The TUNEL assay has been widely applied in different types of cells to detect DNA damage produced by different types of xenobiotics [49, 50].

Kenacid blue binding assay

Through this assay, the change in total protein content is measured, which is a reflection of cell proliferation. If a compound is cytotoxic to the cells, it must affect at least one or more processes involved in cell proliferation, such as DNA synthesis, the proper functioning of organelles such as mitochondria and lysosomes or affect the integrity of the membrane or protein synthesis [51]. When the cell growth is affected, the number of cells present in the treated culture must be reduced with respect to the control, such that the measurement of the concentration of proteins present in the culture constitutes an index of toxicity [51]. Generally, the cells are exposed to the product for evaluation for a time period of 72 h, and the product is then removed and the cells are exposed to the dye, which then binds to the cellular proteins. Finally, the amount of kenacid blue retained by the cells is determined and the percent of inhibition of cell growth is quantified [51, 52].

Resazurin binding assay

Resazurin (7-hydroxy-3H-phenoxazin-3-one 10-oxide) is a non-fluorescent blue dye that is commonly used for the measurements of cell viability [53]. Resazurin is reduced to resofurin (a highly fluorescent pink dye) by oxidoreductases detected primarily in the mitochondria of viable cells. Resofurin is excreted into the medium, allowing continuous monitoring of the proliferation and/or cytotoxicity of the substances in human cells, animals, bacteria, and even fungi [54]. This dye is not extremely toxic to cells and allows the continuity of studies in the same cells, which saves time and money, especially in the primary cultures where the cells are extremely scarce and valuable. Furthermore, it is sensitive and highly reproducible. It is therefore possible to determine in samples at 530–580 nm as excitation wavelengths and 570–620 nm as emission wavelengths, since this dye has both chromophoric and fluorophore properties [53, 54].

Biomarkers

The term "biomarker" is applied to measure an interaction between the biological system and a chemical, physical, or biological agent, which was evaluated as a functional or physiological response that occurs at the cellular or molecular level and is associated with the probability of the development of a disease [55].

The main objectives with the use of biomarkers in human and environmental toxicology are to measure the exposure to xenobiotic agents that causes diseases and to predict the toxic response that could possibly occur [55]. This approach allowed an increase in the requirement of regulation for the development of drugs, pesticides, and other compounds that can produce adverse effects on the human health in addition to greater impacts on occupational health [55, 56]. Short- and long-term toxicity studies in vitro systems and in experimental animals are very valuable to demonstrate the association of different substances with the appearance of cytotoxicity and the development of mutagenesis, carcinogenesis, and teratogenesis in order to promote early actions for the protection of human health [56].

4. Cytotoxic drugs

The absorption of a drug depends on its physicochemical properties, its formulation, and its route of administration. A drug must cross several semi-permeable cell barriers before reaching the systemic circulation. The cell membranes act as biological barriers that selectively inhibit the flow of drug molecules [57, 58].

Within toxic events that can compromise cell functions by xenobiotic agents such as nanoparticle-based drugs, cellular oxidative stress is an important biological process that must be considered [59]. Oxidative stress is manifested by the production of reactive oxygen species (ROS). ROS are highly reactive, very small molecules that are produced as a result of the presence of an unpaired valence electron shell; they are highly reactive and possess the ability to interact with macromolecules such as lipids, proteins, and DNA. There are various cellular signals that are conveyed through binding to antioxidant response elements or in response to electrophiles, which regulate the expression and coordinate different genes related to their chemoprotective and detoxifying capacities [59, 60].

Determining cell viability is essential when analyzing the efficacy of a new drug or treatment [57]. Not all drugs have the same underlying mechanism or the same level of effect, therefore analyzing how they affect cell health can be a key indicator of whether the drugs may work for a specific intended result. The pharmaceutical industry uses a variety of cytotoxicity tests to screen compositions [57]. Chemicals, drugs, and pesticides all affect human cells in different ways, and these tests can uncover the exact mechanisms of how these xenobiotics work in the human body. Cytotoxicity assays can uncover processes such as the destruction of cell membranes, irreversible binding to receptors, impaired protein synthesis, irreversible binding to receptors, and others [19].

Cytotoxic drugs are preferably used to treat neoplastic diseases, these include DNA alkylating agents [61], antimetabolites [62], and microtubule-active agents [63], topoisomerase inhibitors [64], among others. The goal of the pharmaceutical industry is to create increasingly efficient cytotoxic drugs for cancer treatment, with specific targets for certain cellular targets. Nevertheless, they are drugs with high toxicity, mainly utilized for hematopoietic, renal, hepatic, digestive, and dermal ailments [65, 66].

5. Medical devices

According to the WHO, medical devices refer to any instrument, device, implement, machine, implant, reagent for *in vitro* use, software, material, or other similar or related article. These devices must undergo rigorous tests to determine their biocompatibility when they come into contact with the body, regardless of its mechanical, physical, and chemical properties [67].

With the continuous development of science and technology, if medical devices are new, they must undergo biocompatibility tests, cytotoxicity, sensitization, intradermal irritation, acute systemic toxicity, and a series of tests before entering a clinical setting to ensure that it is safe for use and effective in humans [68]. These cytotoxicity studies are generally quick, simple, and straightforward, and help eliminate materials that may be harmful to the body and further consider whether they need further analysis and evaluation for their safe and effective use [69].

Polyurethanes (PUs) represent a popular and important part of industrial products that are characterized by good flexibility properties, high impact resistance, and durability; these characteristics make them polymers with multiple applications [70]. Their block copolymer character provides them a wide versatility in terms of adapting their physical properties and compatibility; thus, PUs are interesting for internal uses (*in vivo*), especially in short-term applications, such as in catheters or implants. Similarly, they are interesting for external use applications (*in vitro*), such as controlled drug release systems. The PUs used as biomedical materials must comply with the mechanical properties for the intended application and must be non-toxic, biodegradable according to the function to be fulfilled, and biocompatible [69, 70].

Biodegradable polymers have gained much attention presently in the medical field in the search for new materials for treating health problems that arise due to their attractive physical properties and good biocompatibility [71].

6. Pesticides

Pesticides, especially herbicides that are used routinely in crop production, have been shown to cause detrimental effects on the human health [72]. These compounds are easily absorbed via different routes, such as the gastrointestinal and respiratory tracts and through the skin. Due to their high stability and affinity for adipose tissues, they can be metabolized and stored in the human organs, mainly in the adipose tissues [72]. Several human diseases have been associated with pesticide exposure, including cancer, hypertension, neurodegenerative diseases (Parkinson's), and diabetes [73–76]. Due to the toxicity and extreme persistence of pesticides in the environment, different studies have been conducted on the toxicity of these agents [77].

Glyphosate for example is a broad spectrum post-emergent herbicide used in both agricultural and non-agricultural areas for weed control [78, 79]. Within the investigations in this regard, Nagy et al. [80] compared the cytotoxic and genotoxic potentials of the active ingredient glyphosate using mononuclear white blood cells that were treated with different concentrations of glyphosate and others with three glyphosate-based herbicides and found that glyphosate induces significant cytotoxicity and genotoxicity effects [80].

The cell metabolic activity is an important indicator of cell viability, and succinate dehydrogenase is an enzyme complex found in the inner mitochondrial membrane of eukaryotic cells that can be used to reflect the viability of these cells [81].

In this field, Devi et al. [82] used rotenone and chlorpyrifos to evaluate the interaction of these pesticides with the protein malate dehydrogenase (MDH) and the consequent cytotoxicity induced by these pesticides. The authors found that rotenone and chlorpyrifos bind strongly to MDH, interfering with protein folding and triggering alterations in their secondary structure.

In this sense, it has also been shown that exposure to pesticides can induce oxidative DNA damage, single and double breaks, and adduct formation [83]. Although there are different DNA repair mechanisms for these damages and for the maintenance of cellular integrity, excessive damage or irreparable damage can lead to cell death processes [8, 83, 84].

In fact, it has been described that some pesticides induce cell death through apoptosis signaling pathways in order to maintain cell homeostasis [85, 86]. Pesticide-induced apoptosis can form the basis of several human diseases, such as cancer and neurological diseases [87]. For this reason, studies in this field have been consistently increasing.

7. Radiation

Radiation, according to its energy, can be classified into ionizing and non-ionizing types [88]. In this sense, ionizing radiation corresponds to the radiation of higher energy (shorter wavelength) within the electromagnetic spectrum. These radiations have sufficient energy to remove electrons from the atoms with which they interact to produce ionizations, while the non-ionizing radiations are those

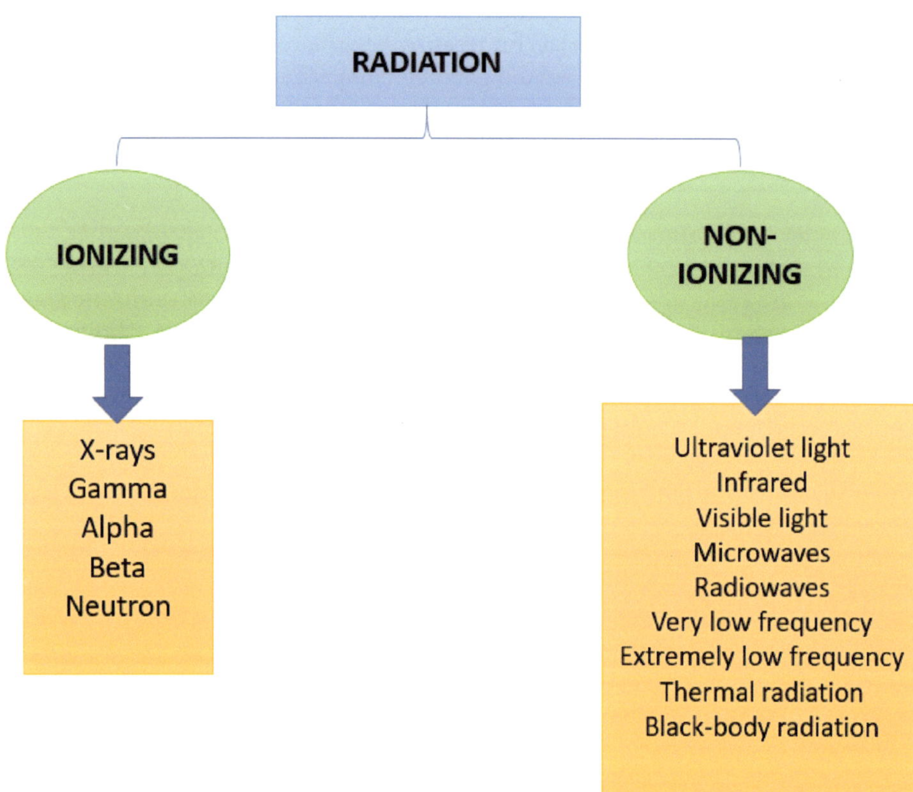

Figure 1.
Types of radiation.

that do not have sufficient energy to remove an electron from the atom they interact with, that is, they do not produce ionizations [88]. **Figure 1** illustrates the types of radiation.

Ionizing radiation is considered to be more dangerous because it has sufficient energy to alter matter (through ionizing energy). Depending on the environment in which the radiation collides, the mechanism of action of the radiation differs. These types of action mechanism can be classified into a direct-action mechanism, which consists of transferring energy to molecules such as proteins, lipids, or DNA, among others, which causes the bond to break down and, consequently, cause damage to the cells. The indirect mechanism of action occurs when energy is absorbed by water molecules in the body and, consequently, radiolysis occurs. In this process, the free radicals of OH^- and the release of H^+ ions are produced, which recombine and can produce H_2, H_2O, and H_2O_2, which when reacting with molecules such as glucose and cholesterol, among others, can cause damage momentary in the metabolism and if they react with DNA, structural damage is generated [89, 90].

These radiations cause DNA damage in different ways, such as double chain breaks, single chain breaks (SSB), hydrogen bond breakage, base dimers, DNA–DNA cross-linking, DNA-protein cross-linking, loss of bases, base modification, and alteration and damage of the repair mechanisms by interaction with cell cycle proteins such as cyclins CDKs and p53 regulator of cell apoptosis, all of which lead to mutations or structural abnormalities that increase the genomic instability [89–91].

8. Conclusions

The damage at the cellular level, either in some structures or in processes that affect cell maintenance, division, or survival, can lead to processes of cell death. An important point in drug evaluation is to provide an alternative approach to improve the predictive capacity of cytotoxicity assays based on cell analysis through incorporation of more specific parameters and/or more appropriate cellular systems. By means of this approach, cytotoxicity can be defined in an integrated manner, based on genomic, proteomic, and cytomic data, starting from the molecules to cells and from the cells to tissues.

The advent of new and improved cytotoxicity assays that are safe, robust, and affordable has been instrumental in advancing the pharmaceutical development process. With these analyzes, presently, research has begun to rely less on animal testing and more on studies that are more economical.

Author details

Grethel León-Mejía[1*], Alvaro Miranda Guevara[1], Ornella Fiorillo Moreno[1] and Carolina Uribe Cruz[2]

1 Facultad de Ciencias Básicas y Biomédicas, Universidad Simón Bolívar, Barranquilla, Colombia

2 Universidad Católica de las Misiones-UCAMI, Posadas, Misiones, Argentina

*Address all correspondence to: gleon4@unisimonbolivar.edu.co

References

[1] Rai, P.K., 2016. Impacts of particulate matter pollution on plants: Implications for environmental biomonitoring. Ecotoxicology and Environmental Safety 129, 120-136. https://doi.org/10.1016/j.ecoenv.2016.03.012

[2] Al-Thani, H., Koç, M., Isaifan, R.J., 2018. A review on the direct effect of particulate atmospheric pollution on materials and its mitigation for sustainable cities and societies. Environ Sci Pollut Res 25, 27839-27857. https://doi.org/10.1007/s11356-018-2952-8

[3] Oesch, F., Fabian, E., Guth, K., Landsiedel, R., 2014. Xenobiotic-metabolizing enzymes in the skin of rat, mouse, pig, guinea pig, man, and in human skin models. Arch Toxicol 88, 2135-2190. https://doi.org/10.1007/s00204-014-1382-8

[4] Tabrez, S., Shakil, S., Urooj, M., Damanhouri, G.A., Abuzenadah, A.M., Ahmad, M., 2011. Genotoxicity Testing and Biomarker Studies on Surface Waters: An Overview of the Techniques and Their Efficacies. Journal of Environmental Science and Health, Part C 29, 250-275. https://doi.org/10.1080/10590501.2011.601849

[5] Mukherjee, P.K., 2019. Bioassay-Guided Isolation and Evaluation of Herbal Drugs, in: Quality Control and Evaluation of Herbal Drugs. Elsevier, pp. 515-537. https://doi.org/10.1016/B978-0-12-813374-3.00013-2

[6] Pagano, M., Faggio, C., 2015. The use of erythrocyte fragility to assess xenobiotic cytotoxicity: Cell Death and Erythrocytes. Cell Biochem Funct 33, 351-355. https://doi.org/10.1002/cbf.3135

[7] Yan, J.S., Chen, M., Wang, Q., 2013. New Drug Regulation and Approval in China, in: A Comprehensive Guide to Toxicology in Preclinical Drug Development. Elsevier, pp. 713-723. https://doi.org/10.1016/B978-0-12-387815-1.00029-0

[8] Jarvis, I.W.H., Dreij, K., Mattsson, Å., Jernström, B., Stenius, U., 2014. Interactions between polycyclic aromatic hydrocarbons in complex mixtures and implications for cancer risk assessment. Toxicology 321, 27-39. https://doi.org/10.1016/j.tox.2014.03.012

[9] De Zio, D., Cianfanelli, V., Cecconi, F., 2013. New Insights into the Link Between DNA Damage and Apoptosis. Antioxidants & Redox Signaling 19, 559-571. https://doi.org/10.1089/ars.2012.4938

[10] Wnuk, A., Kajta, M., 2017. Steroid and Xenobiotic Receptor Signalling in Apoptosis and Autophagy of the Nervous System. IJMS 18, 2394. https://doi.org/10.3390/ijms18112394

[11] D'Arcy, M.S., 2019. Cell death: a review of the major forms of apoptosis, necrosis and autophagy. Cell Biol Int 43, 582-592. https://doi.org/10.1002/cbin.11137

[12] Wyllie, A.H., Kerr, J.F.R., Currie, A.R., 1980. Cell Death: The Significance of Apoptosis, in: International Review of Cytology. Elsevier, pp. 251-306. https://doi.org/10.1016/S0074-7696(08)62312-8

[13] Elmore, S., 2007. Apoptosis: A Review of Programmed Cell Death. Toxicol Pathol 35, 495-516. https://doi.org/10.1080/01926230701320337

[14] Jeong, S.-Y., Seol, D.-W., 2008. The role of mitochondria in apoptosis. BMB Reports 41, 11-22. https://doi.org/10.5483/BMBRep.2008.41.1.011

[15] Mukhopadhyay, S., Panda, P.K., Sinha, N., Das, D.N., Bhutia, S.K., 2014. Autophagy and apoptosis:

where do they meet? Apoptosis 19, 555-566. https://doi.org/10.1007/s10495-014-0967-2

[16] Saha, S., Panigrahi, D.P., Patil, S., Bhutia, S.K., 2018. Autophagy in health and disease: A comprehensive review. Biomedicine & Pharmacotherapy 104, 485-495. https://doi.org/10.1016/j.biopha.2018.05.007

[17] Glick, D., Barth, S., Macleod, K.F., 2010. Autophagy: cellular and molecular mechanisms. J. Pathol. 221, 3-12. https://doi.org/10.1002/path.2697

[18] Levy, J.M.M., Towers, C.G., Thorburn, A., 2017. Targeting autophagy in cancer. Nat Rev Cancer 17, 528-542. https://doi.org/10.1038/nrc.2017.53

[19] Adan, A., Kiraz, Y., Baran, Y., 2016. Cell Proliferation and Cytotoxicity Assays. CPB 17, 1213-1221. https://doi.org/10.2174/1389201017666160808160513

[20] Caldas, I.P., Alves, G.G., Barbosa, I.B., Scelza, P., de Noronha, F., Scelza, M.Z., 2019. In vitro cytotoxicity of dental adhesives: A systematic review. Dental Materials 35, 195-205. https://doi.org/10.1016/j.dental.2018.11.028

[21] Ilboudo, S., Fouche, E., Rizzati, V., Toé, A.M., Gamet-Payrastre, L., Guissou, P.I., 2014. In vitro impact of five pesticides alone or in combination on human intestinal cell line Caco-2. Toxicology Reports 1, 474-489. https://doi.org/10.1016/j.toxrep.2014.07.008

[22] Venkatesan, R., Park, Y.U., Ji, E., Yeo, E.-J., Kim, S.Y., 2017. Malathion increases apoptotic cell death by inducing lysosomal membrane permeabilization in N2a neuroblastoma cells: a model for neurodegeneration in Alzheimer's disease. Cell Death Discov. 3, 17007. https://doi.org/10.1038/cddiscovery.2017.7

[23] Yuan, G.-J., Deng, J.-J., Cao, D.-D., Shi, L., Chen, X., Lei, J.-J., Xu, X.-M., 2017. Autophagic cell death induced by reactive oxygen species is involved in hyperthermic sensitization to ionizing radiation in human hepatocellular carcinoma cells. WJG 23, 5530. https://doi.org/10.3748/wjg.v23.i30.5530

[24] Fazel, M., Mehnati, P., Baradaran, B., Pirayesh, J., 2016. Evaluation of gamma radiation-induced cytotoxicity of breast cancer cells: Is there a time-dependent dose with high efficiency? Indian J Cancer 53, 25. https://doi.org/10.4103/0019-509X.180862

[25] Hardneck, F., Israel, G., Pool, E., Maree, L., 2018. Quantitative assessment of heavy metal effects on sperm function using computer-aided sperm analysis and cytotoxicity assays. Andrologia 50, e13141. https://doi.org/10.1111/and.13141

[26] Karri, V., Kumar, V., Ramos, D., Oliveira, E., Schuhmacher, M., 2018. An in vitro cytotoxic approach to assess the toxicity of heavy metals and their binary mixtures on hippocampal HT-22 cell line. Toxicology Letters 282, 25-36. https://doi.org/10.1016/j.toxlet.2017.10.002

[27] Yadav, A., Mandal, M.K., Dubey, K.K., 2020. In Vitro Cytotoxicity Study of Cyclophosphamide, Etoposide and Paclitaxel on Monocyte Macrophage Cell Line Raw 264.7. Indian J Microbiol 60, 511-517. https://doi.org/10.1007/s12088-020-00896-1

[28] Florento, L., Matias, R., Tuaño, E., Santiago, K., Dela Cruz, F., Tuazon, A., 2012. Comparison of Cytotoxic Activity of Anticancer Drugs against Various Human Tumor Cell Lines Using In Vitro Cell-Based Approach. Int J Biomed Sci 8, 76-80.

[29] Sharma, A., Gorey, B., Casey, A., 2019. *In vitro* comparative cytotoxicity study of aminated polystyrene, zinc

oxide and silver nanoparticles on a cervical cancer cell line. Drug and Chemical Toxicology 42, 9-23. https://doi.org/10.1080/01480545.2018.1424181

[30] Adeyemi, J.A., Machado, A.R.T., Ogunjimi, A.T., Alberici, L.C., Antunes, L.M.G., Barbosa, F., 2020. Cytotoxicity, mutagenicity, oxidative stress and mitochondrial impairment in human hepatoma (HepG2) cells exposed to copper oxide, copper-iron oxide and carbon nanoparticles. Ecotoxicology and Environmental Safety 189, 109982. https://doi.org/10.1016/j.ecoenv.2019.109982

[31] Wawrzynkiewicz, A., Rozpedek-Kaminska, W., Galita, G., Lukomska-Szymanska, M., Lapinska, B., Sokolowski, J., Majsterek, I., 2020. The Cytotoxicity and Genotoxicity of Three Dental Universal Adhesives—An In Vitro Study. IJMS 21, 3950. https://doi.org/10.3390/ijms21113950

[32] Romano, M.R., Ferrara, M., Gatto, C., Ferrari, B., Giurgola, L., D'Amato Tóthová, J., 2019. Evaluation of Cytotoxicity of Perfluorocarbons for Intraocular Use by Cytotoxicity Test In Vitro in Cell Lines and Human Donor Retina Ex Vivo. Trans. Vis. Sci. Tech. 8, 24. https://doi.org/10.1167/tvst.8.5.24

[33] Xu, C., Chen, G., Tan, Z., Hu, Z., Qu, Z., Zhang, Q., Lu, Maoping, Wu, K., Lu, Mangeng, Liang, L., 2020. Evaluation of cytotoxicity in vitro and properties of polysiloxane-based polyurethane/lignin elastomers. Reactive and Functional Polymers 149, 104514. https://doi.org/10.1016/j.reactfunctpolym.2020.104514

[34] Ates, G., Vanhaecke, T., Rogiers, V., Rodrigues, R.M., 2017. Assaying Cellular Viability Using the Neutral Red Uptake Assay, in: Gilbert, D.F., Friedrich, O. (Eds.), Cell Viability Assays, Methods in Molecular Biology. Springer New York, New York, NY, pp. 19-26. https://doi.org/10.1007/978-1-4939-6960-9_2

[35] Mello, D.F., Trevisan, R., Rivera, N., Geitner, N.K., Di Giulio, R.T., Wiesner, M.R., Hsu-Kim, H., Meyer, J.N., 2020. Caveats to the use of MTT, neutral red, Hoechst and Resazurin to measure silver nanoparticle cytotoxicity. Chemico-Biological Interactions 315, 108868. https://doi.org/10.1016/j.cbi.2019.108868

[36] Pascua-Maestro, R., Corraliza-Gomez, M., Diez-Hermano, S., Perez-Segurado, C., Ganfornina, M.D., Sanchez, D., 2018. The MTT-formazan assay: Complementary technical approaches and in vivo validation in Drosophila larvae. Acta Histochemica 120, 179-186. https://doi.org/10.1016/j.acthis.2018.01.006

[37] van Meerloo, J., Kaspers, G.J.L., Cloos, J., 2011. Cell Sensitivity Assays: The MTT Assay, in: Cree, I.A. (Ed.), Cancer Cell Culture, Methods in Molecular Biology. Humana Press, Totowa, NJ, pp. 237-245. https://doi.org/10.1007/978-1-61779-080-5_20

[38] Sylvester, P.W., 2011. Optimization of the Tetrazolium Dye (MTT) Colorimetric Assay for Cellular Growth and Viability, in: Satyanarayanajois, S.D. (Ed.), Drug Design and Discovery, Methods in Molecular Biology. Humana Press, Totowa, NJ, pp. 157-168. https://doi.org/10.1007/978-1-61779-012-6_9

[39] Rafehi, H., Orlowski, C., Georgiadis, G.T., Ververis, K., El-Osta, A., Karagiannis, T.C., 2011. Clonogenic Assay: Adherent Cells. JoVE 2573. https://doi.org/10.3791/2573

[40] Nuryadi, E., Mayang Permata, T.B., Komatsu, S., Oike, T., Nakano, T., 2018. Inter-assay precision of clonogenic assays for radiosensitivity in cancer cell line A549. Oncotarget 9, 13706-13712. https://doi.org/10.18632/oncotarget.24448

[41] Orellana, E., Kasinski, A., 2016. Sulforhodamine B (SRB) Assay in Cell

Culture to Investigate Cell Proliferation. BIO-PROTOCOL 6. https://doi.org/10.21769/BioProtoc.1984

[42] Vichai, V., Kirtikara, K., 2006. Sulforhodamine B colorimetric assay for cytotoxicity screening. Nat Protoc 1, 1112-1116. https://doi.org/10.1038/nprot.2006.179

[43] Parhamifar, L., Andersen, H., Moghimi, S.M., 2019. Lactate Dehydrogenase Assay for Assessment of Polycation Cytotoxicity, in: Ogris, M., Sami, H. (Eds.), Nanotechnology for Nucleic Acid Delivery, Methods in Molecular Biology. Springer New York, New York, NY, pp. 291-299. https://doi.org/10.1007/978-1-4939-9092-4_18

[44] Kumar, P., Nagarajan, A., Uchil, P.D., 2018. Analysis of Cell Viability by the Lactate Dehydrogenase Assay. Cold Spring Harb Protoc 2018, pdb.prot095497. https://doi.org/10.1101/pdb.prot095497

[45] Crowley, L.C., Marfell, B.J., Scott, A.P., Waterhouse, N.J., 2016a. Quantitation of Apoptosis and Necrosis by Annexin V Binding, Propidium Iodide Uptake, and Flow Cytometry. Cold Spring Harb Protoc 2016, pdb.prot087288. https://doi.org/10.1101/pdb.prot087288

[46] Duensing, T.D., Watson, S.R., 2018. Assessment of Apoptosis (Programmed Cell Death) by Flow Cytometry. Cold Spring Harb Protoc 2018, pdb.prot093807. https://doi.org/10.1101/pdb.prot093807

[47] Crowley, L.C., Marfell, B.J., Waterhouse, N.J., 2016b. Detection of DNA Fragmentation in Apoptotic Cells by TUNEL. Cold Spring Harb Protoc 2016, pdb.prot087221. https://doi.org/10.1101/pdb.prot087221

[48] Kyrylkova, K., Kyryachenko, S., Leid, M., Kioussi, C., 2012. Detection of Apoptosis by TUNEL Assay, in: Kioussi, C. (Ed.), Odontogenesis, Methods in Molecular Biology. Humana Press, Totowa, NJ, pp. 41-47. https://doi.org/10.1007/978-1-61779-860-3_5

[49] Li, C., Zhao, K., Zhang, H., Liu, L., Xiong, F., Wang, K., Chen, B., 2018. Lead exposure reduces sperm quality and DNA integrity in mice. Environmental Toxicology 33, 594-602. https://doi.org/10.1002/tox.22545

[50] Yang, L., Tu, D., Zhao, Z., Cui, J., 2017. Cytotoxicity and apoptosis induced by mixed mycotoxins (T-2 and HT-2 toxin) on primary hepatocytes of broilers in vitro. Toxicon 129, 1-10. https://doi.org/10.1016/j.toxicon.2017.01.001

[51] Clothier, R., Gottschalg, E., Casati, S., Balls, M., 2006. The FRAME Alternatives Laboratory Database. 1. *In Vitro* Basal Cytotoxicity determined by the Kenacid Blue Total Protein Assay. Altern Lab Anim 34, 151-175. https://doi.org/10.1177/026119290603400203

[52] Hurst, H.S., Clothier, R.H., Pratten, M., 2007. An Evaluation of a Novel Chick Cardiomyocyte Micromass Culture Assay with Two Teratogens/Embryotoxins Associated with Heart Defects. Altern Lab Anim 35, 505-514. https://doi.org/10.1177/026119290703500510

[53] Csepregi, R., Lemli, B., Kunsági-Máté, S., Szente, L., Kőszegi, T., Németi, B., Poór, M., 2018. Complex Formation of Resorufin and Resazurin with Β-Cyclodextrins: Can Cyclodextrins Interfere with a Resazurin Cell Viability Assay? Molecules 23, 382. https://doi.org/10.3390/molecules23020382

[54] Silva, F.S.G., Starostina, I.G., Ivanova, V.V., Rizvanov, A.A., Oliveira, P.J., Pereira, S.P., 2016. Determination of Metabolic Viability and Cell Mass Using a Tandem Resazurin/Sulforhodamine B Assay: SRB/Resazurin Assay for Cell

Viability and Metabolism. Current Protocols in Toxicology 68, 2.24.1-2.24.15. https://doi.org/10.1002/cptx.1

[55] Santonen, T., Aitio, A., Fowler, B.A., Nordberg, M., 2015. Biological Monitoring and Biomarkers, in: Handbook on the Toxicology of Metals. Elsevier, pp. 155-171. https://doi.org/10.1016/B978-0-444-59453-2.00008-1

[56] Aronson, J.K., Ferner, R.E., 2017. Biomarkers—A General Review. Current Protocols in Pharmacology 76. https://doi.org/10.1002/cpph.19

[57] Hilde Bohets, Pieter Annaert, Geert Mannens, Ludy van Beijsterveldt, Katelijne Anciaux, Peter Verboven, Willem Meuldermans, Karel Lavrijsen, 2001. Strategies for Absorption Screening in Drug Discovery and Development. CTMC 1, 367-383. https://doi.org/10.2174/1568026013394886

[58] Lee, Y., Choi, S.Q., 2019. Quantitative analysis for lipophilic drug transport through a model lipid membrane with membrane retention. European Journal of Pharmaceutical Sciences 134, 176-184. https://doi.org/10.1016/j.ejps.2019.04.020

[59] Yang, B., Chen, Y., Shi, J., 2019. Reactive Oxygen Species (ROS)-Based Nanomedicine. Chem. Rev. 119, 4881-4985. https://doi.org/10.1021/acs.chemrev.8b00626

[60] Abdal Dayem, A., Hossain, M., Lee, S., Kim, K., Saha, S., Yang, G.-M., Choi, H., Cho, S.-G., 2017. The Role of Reactive Oxygen Species (ROS) in the Biological Activities of Metallic Nanoparticles. IJMS 18, 120. https://doi.org/10.3390/ijms18010120

[61] Cheng, S.-Y., Seo, J., Huang, B.T., Napolitano, T., Champeil, E., 2016. Mitomycin C and decarbamoyl mitomycin C induce p53-independent p21WAF1/CIP1 activation. International Journal of Oncology 49, 1815-1824. https://doi.org/10.3892/ijo.2016.3703

[62] Petaccia, M., Gentili, P., Bešker, N., D'Abramo, M., Giansanti, L., Leonelli, F., La Bella, A., Gradella Villalva, D., Mancini, G., 2016. Kinetics and mechanistic study of competitive inhibition of thymidine phosphorylase by 5-fluoruracil derivatives. Colloids and Surfaces B: Biointerfaces 140, 121-127. https://doi.org/10.1016/j.colsurfb.2015.12.020

[63] Wei, C., Pan, Y., Huang, H., Li, Y.-P., 2018. Estramustine phosphate induces prostate cancer cell line PC3 apoptosis by down-regulating miR-31 levels. European Review for Medical and Pharmacological Sciences 22, 40-45. https://doi.org/10.26355/eurrev_201801_14098

[64] Kovács E.R., Tóth S., Erdélyi D.J., 2018. Az etopozid szervezeten belüli eloszlásában és metabolizmusában szerepet játszó epigenetikai hatások. Orvosi Hetilap 159, 1295-1302. https://doi.org/10.1556/650.2018.31162

[65] Kitao, H., Iimori, M., Kataoka, Y., Wakasa, T., Tokunaga, E., Saeki, H., Oki, E., Maehara, Y., 2018. DNA replication stress and cancer chemotherapy. Cancer Sci 109, 264-271. https://doi.org/10.1111/cas.13455

[66] Lin, A., Giuliano, C.J., Palladino, A., John, K.M., Abramowicz, C., Yuan, M.L., Sausville, E.L., Lukow, D.A., Liu, L., Chait, A.R., Galluzzo, Z.C., Tucker, C., Sheltzer, J.M., 2019. Off-target toxicity is a common mechanism of action of cancer drugs undergoing clinical trials. Sci. Transl. Med. 11, eaaw8412. https://doi.org/10.1126/scitranslmed.aaw8412

[67] Who - World Health Organization, 2017, ISBN 9789241512350 https://www.who.int/health-topics/medical-devices#tab=tab_1

[68] Reeve, L., Baldrick, P., 2017. Biocompatibility assessments for medical devices – evolving regulatory considerations. Expert Review of Medical Devices 14, 161-167. https://doi.org/10.1080/17434440.2017.1280392

[69] Bernard, M., Jubeli, E., Pungente, M.D., Yagoubi, N., 2018. Biocompatibility of polymer-based biomaterials and medical devices – regulations, *in vitro* screening and risk-management. Biomater. Sci. 6, 2025-2053. https://doi.org/10.1039/C8BM00518D

[70] Zuber, M., Zia, F., Zia, K.M., Tabasum, S., Salman, M., Sultan, N., 2015. Collagen based polyurethanes—A review of recent advances and perspective. International Journal of Biological Macromolecules 80, 366-374. https://doi.org/10.1016/j.ijbiomac.2015.07.001

[71] Kenry, Liu, B., 2018. Recent Advances in Biodegradable Conducting Polymers and Their Biomedical Applications. Biomacromolecules 19, 1783-1803. https://doi.org/10.1021/acs.biomac.8b00275

[72] Jabłońska-Trypuć, A., Wydro, U., Wołejko, E., Butarewicz, A., 2019. Toxicological Effects of Traumatic Acid and Selected Herbicides on Human Breast Cancer Cells: In Vitro Cytotoxicity Assessment of Analyzed Compounds. Molecules 24, 1710. https://doi.org/10.3390/molecules24091710

[73] Jemal, A., Bray, F., Center, M.M., Ferlay, J., Ward, E., Forman, D., 2011. Global cancer statistics. CA: A Cancer Journal for Clinicians 61, 69-90. https://doi.org/10.3322/caac.20107

[74] Thapa, K., Pant, B.R., 2014. Pesticides in vegetable and food commodities: environment and public health concern. J Nepal Health Res Counc 12, 208-210.

[75] Ball, N., Teo, W.-P., Chandra, S., Chapman, J., 2019. Parkinson's Disease and the Environment. Front. Neurol. 10, 218. https://doi.org/10.3389/fneur.2019.00218

[76] Brust, R.S., Oliveira, L.P.M. de, Silva, A.C.S.S. da, Regazzi, I.C.R., Aguiar, G.S. de, Knupp, V.M. de A.O., 2019. Epidemiological profile of farmworkers from the state of Rio de Janeiro. Rev. Bras. Enferm. 72, 122-128. https://doi.org/10.1590/0034-7167-2017-0555

[77] Bilal, M., Iqbal, H.M.N., Barceló, D., 2019. Persistence of pesticides-based contaminants in the environment and their effective degradation using laccase-assisted biocatalytic systems. Science of The Total Environment 695, 133896. https://doi.org/10.1016/j.scitotenv.2019.133896

[78] Chikoye, D., Lum, A.F., Udensi, U.E., 2010. Efficacy of a new glyphosate formulation for weed control in maize in southwest Nigeria. Crop Protection 29, 947-952. https://doi.org/10.1016/j.cropro.2010.06.011

[79] Uren Webster, T.M., Laing, L.V., Florance, H., Santos, E.M., 2014. Effects of Glyphosate and its Formulation, Roundup, on Reproduction in Zebrafish (Danio rerio). Environ. Sci. Technol. 48, 1271-1279. https://doi.org/10.1021/es404258h

[80] Nagy, K., Tessema, R.A., Budnik, L.T., Ádám, B., 2019. Comparative cyto- and genotoxicity assessment of glyphosate and glyphosate-based herbicides in human peripheral white blood cells. Environmental Research 179, 108851. https://doi.org/10.1016/j.envres.2019.108851

[81] Stockert, J.C., Horobin, R.W., Colombo, L.L., Blázquez-Castro, A., 2018. Tetrazolium salts and formazan products in Cell Biology: Viability assessment, fluorescence imaging,

and labeling perspectives. Acta Histochemica 120, 159-167. https://doi.org/10.1016/j.acthis.2018.02.005

[82] Devi, S., Karsauliya, K., Srivastava, T., Raj, R., Kumar, D., Priya, S., 2021. Pesticide interactions induce alterations in secondary structure of malate dehydrogenase to cause destability and cytotoxicity. Chemosphere 263, 128074. https://doi.org/10.1016/j.chemosphere.2020.128074

[83] Benedetti, D., Lopes Alderete, B., de Souza, C.T., Ferraz Dias, J., Niekraszewicz, L., Cappetta, M., Martínez-López, W., Da Silva, J., 2018. DNA damage and epigenetic alteration in soybean farmers exposed to complex mixture of pesticides. Mutagenesis 33, 87-95. https://doi.org/10.1093/mutage/gex035

[84] Kapka-Skrzypczak, L., Czajka, M., Sawicki, K., Matysiak-Kucharek, M., Gabelova, A., Sramkova, M., Bartyzel-Lechforowicz, H., Kruszewski, M., 2019. Assessment of DNA damage in Polish children environmentally exposed to pesticides. Mutation Research/Genetic Toxicology and Environmental Mutagenesis 843, 52-56. https://doi.org/10.1016/j.mrgentox.2018.12.012

[85] Jiang, J., Wu, S., Liu, X., Wang, Y., An, X., Cai, L., Zhao, X., 2015. Effect of acetochlor on transcription of genes associated with oxidative stress, apoptosis, immunotoxicity and endocrine disruption in the early life stage of zebrafish. Environmental Toxicology and Pharmacology 40, 516-523. https://doi.org/10.1016/j.etap.2015.08.005

[86] Zerin, T., Song, H.-Y., Kim, Y.-S., 2015. Extracellular signal-regulated kinase pathway play distinct role in acetochlor-mediated toxicity and intrinsic apoptosis in A549 cells. Toxicology in Vitro 29, 85-92. https://doi.org/10.1016/j.tiv.2014.09.011

[87] Mattson, M.P., 2000. Apoptosis in neurodegenerative disorders. Nat Rev Mol Cell Biol 1, 120-130. https://doi.org/10.1038/35040009

[88] Hamada, N., Fujimichi, Y., 2014. Classification of radiation effects for dose limitation purposes: history, current situation and future prospects. Journal of Radiation Research 55, 629-640. https://doi.org/10.1093/jrr/rru019

[89] Rak, J., Chomicz, L., Wiczk, J., Westphal, K., Zdrowowicz, M., Wityk, P., Żyndul, M., Makurat, S., Golon, Ł., 2015. Mechanisms of Damage to DNA Labeled with Electrophilic Nucleobases Induced by Ionizing or UV Radiation. J. Phys. Chem. B 119, 8227-8238. https://doi.org/10.1021/acs.jpcb.5b03948

[90] Runge, R., Arlt, J., Oehme, L., Freudenberg, R., Kotzerke, J., 2017. Comparison of clonogenic cell survival and DNA damage induced by 188Re and X-rays in rat thyroid cells. Nuklearmedizin 56, 47-54. https://doi.org/10.3413/Nukmed-0842-16-08

[91] Zhang, X., Ye, C., Sun, F., Wei, W., Hu, B., Wang, J., 2016. Both Complexity and Location of DNA Damage Contribute to Cellular Senescence Induced by Ionizing Radiation. PLoS ONE 11, e0155725. https://doi.org/10.1371/journal.pone.0155725

Chapter 3

Cytotoxic and Antimicrobial Activities of Quinones Isolated from Different Organism

Nimsi Campos-Xolalpa, Julia Pérez-Ramos,
Ana Esquivel-Campos, Cuauhtemoc Pérez-González,
Leonor Sánchez-Pérez and Salud Pérez-Gutiérrez

Abstract

Cancer is a group of related diseases in which there is uncontrolled cell growth that spreads to the surrounding tissues and damages them. Cancer remains the disease with the leading cause of death worldwide, and incidence and mortality are increasing rapidly. The main cancer treatment is chemotherapy; however, the compounds used in this treatment have serious side effects for this reason, is necessary to develop new therapeutic strategies. Natural products are an excellent pharmacological alternative for the treatment of cancer and infections. In search of new compounds with cytotoxic and antimicrobial activity, we have found quinones that have a high pharmacological potency in the treatment of these health problems. Quinones are an aromatic system of one or diketone and are mainly isolated from plants, fungi, bacteria, and other organisms. These compounds are secondary metabolites derived from the oxidation of hydroquinones; they include benzoquinones, naphthoquinones, anthraquinones, and polyquinones. This review summarizes the activity of 152 anticancer and 30 antimicrobial quinones.

Keywords: quinones, cancer, cytotoxic, antimicrobial, natural product

1. Introduction

Cancer is a group of a collection of related diseases where there is uncontrolled cell growth and spread into surrounding tissues, producing damage to them. In many cases, these cells form tumors and some cancer cells travel through the lymphatic system or blood to other places of the body and form new tumors.

Cancer remains the disease with major cause of death globally. In 2018, there were reported about 18 million new cases of cancer [1] and approximately 9.6 million deaths from this disease [2]; in addition, all over the world, the incidence and mortality of cancer are increasing. The risk of incidence of cancer is associated with age, infections, and human habits like poor diet, consumption of alcohol, tobacco, and others [3]; also, there are genetic predisposition and immune conditions [4].

Diseases due to the infections of bacteria and fungi are a very important health problem throughout the world. In 2019, the incidence of infection transmitted by food and water increased. The treatment of infection by bacteria is the administration of antibiotics; however, these drugs have been losing effectiveness because there is increased bacterial drug resistance [5]. The main causes of bacterial resistance are unnecessary prescriptions [6] and the unregulated antibiotics sale in many countries, leading to inadequate and unnecessary consumption [7]. Then, infections treatments become more expensive and have less effective.

From ancient times, natural products have been used in the treatment of different diseases, for example, in Egypt around 1550 BC, the "Ebers Papyrus" reported the use of 700 drugs [8]. Nowadays, natural products are an important source of compounds with great potential for the treatment of infections and different forms of cancer [9].

Quinones are an important family of natural products. They have a variety of biological effects, such as anticancer and antimicrobial activities [10, 11]. The 1,4-naphthoquinones, since ancient times, have been used as cosmetics for coloring skin, as well as the treatment of some diseases. These compounds have several activities like anti-inflammatory, antiviral, anticancer, and antibacterial, among others.

For example, juglone and plumbagin show an antimicrobial effect on bacteria and fungi, and they are defensive compounds in the plant. Cytosporaquine A-D and physcion exhibited cytotoxic activity against several human cell cancer lines.

The cytotoxic and antimicrobial activities of 1,4 naphthoquinones are due mainly to two carbonyl groups present in these compounds, which can accept one or two electrons to form a semiquinone radical or di-anion species and for their acid–base properties [10].

The present review focuses on the anticancer and antimicrobial activities of 182 quinones isolated from natural sources in the last 5 years (**Tables 1** and **2**).

2. Anticancer and antimicrobial activity of quinones obtained from plants, animals, and microorganisms

The incidence of cancer has increased; in 2018, around 9.6 million deaths in the world were due to this disease. The drugs used in chemotherapy have side effects and the cancer cells can have resistance to these drugs. Therefore, the study of new molecules with anticancer activity has become important.

Infectious diseases are an international health public problem, especially in undeveloped countries. For the treatment of these diseases are used antibiotics; however, several microorganisms present resistance to these drugs.

The search for new compounds with these activities has become important. Plants, marine organisms, fungi, and bacteria are natural sources to obtain substances with pharmacological effects.

Quinones are natural products with different pharmacological activities, such as anticancer and antimicrobial effects. These compounds can be obtained by synthesis or the structure modified to increase their activity.

This chapter shows the revision of the literature generated in the last 5 years of quinones isolated from 65 plant species, bacteria, fungi, algae, or sponges. The plants were the most different species studied, followed by fungi with 10 species, Streptomyces with 4 strain investigated, and bacteria with only one studied. Nowadays, the study of marine organisms has become more important, with 3 species of sponges studied and from which these compounds have been isolated, and there was 1 scorpion studied.

The cytotoxic properties of isolated quinones in the period 2015 to 2020 were mainly determined by in vitro and in vivo studies. This was due to some factors such as the sensitivity of these tests and the consumption of small amounts of

Quinona/ Natural origin (plants, animals, microorganisms)	Structure	Activity or model where it was tested	Results	Ref.
Plants				
7-(3′,4′-dihydroxy-benzene)-2,3-dimethoxy-1,4-naphthoquinone. (ajaniquinone). *Ajania salicifolia*	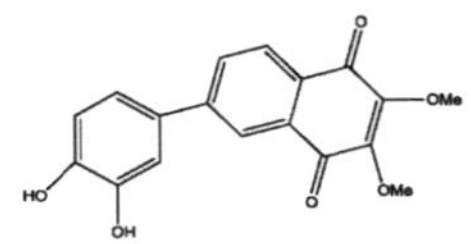	MTT assay Hela HepG2 K562	IC_{50} (μM) 19.68 28.71 13.75	[12]
2-hydroxy-5-ethoxy-3-nonyl-1,4-benzoquinone (1). 5-*O*-butyl-embelin (2). 5-O-methylembelin (3). 5-O-methyl-rapanone (4). 5-O-ethylembelin (5). *Aegiceras corniculatum*	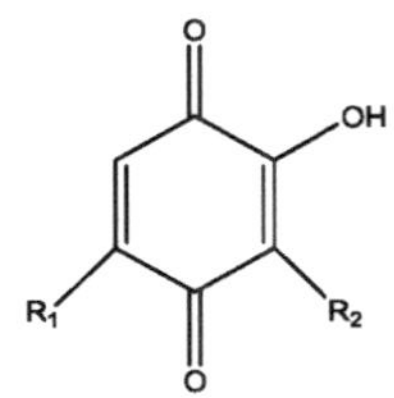 R1 R₂ 1) $OCCH_2CH_3$ C_9H_{19} 2) $OC(CH_2)2CH_3$ C11H_{23} 3) OCH_3 $C_{11}H_{23}$ 4) OCH_3 $C_{13}H_{27}$ 5) $OCH2CH_3$ $C_{11}H_{23}$	MTT assay	IC_{50} μM HL-60 HepG2 BGC-823 A2780 (1) 18 48.2 24 20 (2) 8.77 38.6 9.70 14.48 (3) 8.79 43.08 10.63 15.60 (4) 7.60 40.10 10.40 14.50 (5) 11.65 > 100 13.07 10.58	[13]
Aloesaponarin II (1). Aloesaponarin I (2). *Aloe megalacantha*	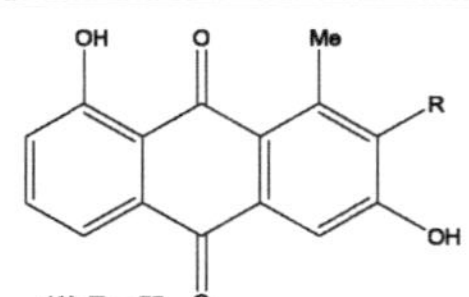 (1) R= H (2) R= COOMe	CAF KB-3-1	IC_{50} μM (1) 0.98 (2) 16	[14]

Quinona/ Natural origin (plants, animals, microorganisms)	Structure	Activity or model where it was tested	Results	Ref.
Ardisiaquinone A (1) Ardisiaquinone B (2) Ardisiaquinone C (3) Ardisiaquinone D (4) Ardisiaquinone E (5) Ardisiaquinone F (6) Ardisiaquinone G (7) Ardisiaquinone H (8) *Ardisia quinquegona*	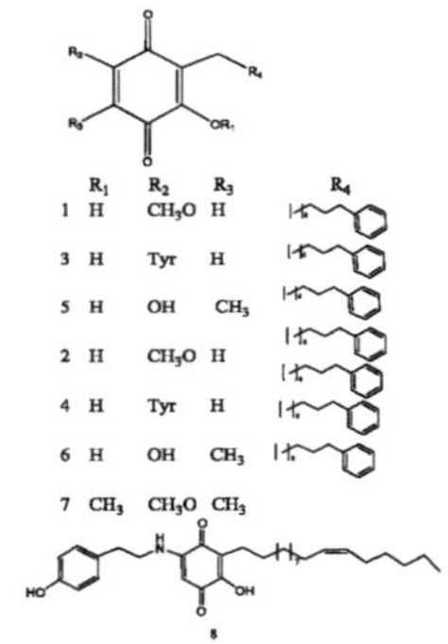	MTT assay A549	IC_{50} µM (1) 55.7 (2) 44.2 (3) 67.9 (4) 75.6 (5) 12.7 (6) 15.2 (7) 47.5 (8) 67.0	[15]
Isovalerylalkannin (1) α-methyl-n-butyl alkannin (2) Acetylalkanni (3) β-acetoxy isovalerylalkannin (4) Alkannin (5) 4-hydroxy 4-methyl valeryl alkannin (6) *Arnebia densiflora*	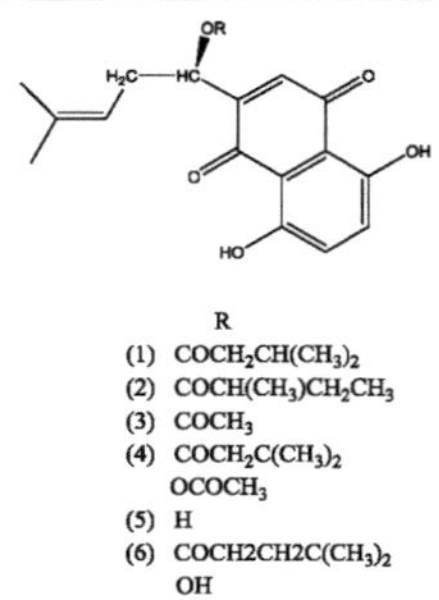	MTT assay L929 HeLa HEp-2	IC_{50} µg/mL Range 26.34–172.35	[16]
Benzoquinone *Artemisia asiática*	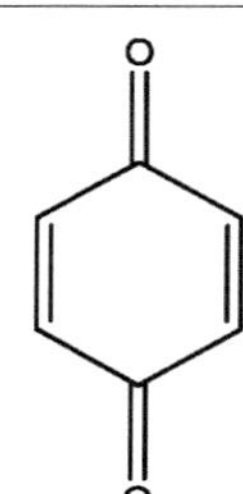	MTT assay A431 SYF	IC_{50} µM 54.1 52.3	[17]

DOI: http://dx.doi.org/10.5772/intechopen.95598

Quinona/ Natural origin (plants, animals, microorganisms)	Structure	Activity or model where it was tested	Results	Ref.
Acetylshikonin *Lithospermum erythrorhizon* *Onosma visianii*	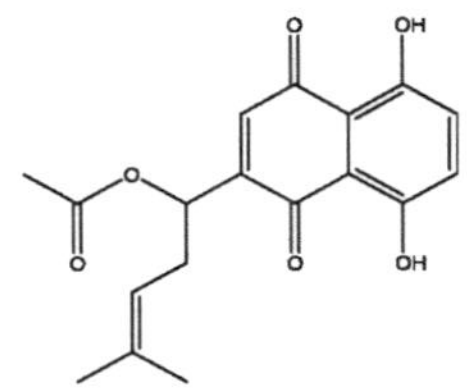	WST-1 cell viability assay HepG2 MTT assay MDA-MB231 4 T1 MDA-MB231 HCT-116	IC_{50} μM 2 IC_{50} μM 24 h 48 h 72 h 9.11 3.34 1.83 4.98 2.61 1.74 IC_{50} μg/mL 72 h 80.2 24.6	[18] [19] [20]
Shikonin *Lithospermum erythrorhizon.* Different species of *Boraginaceae* family	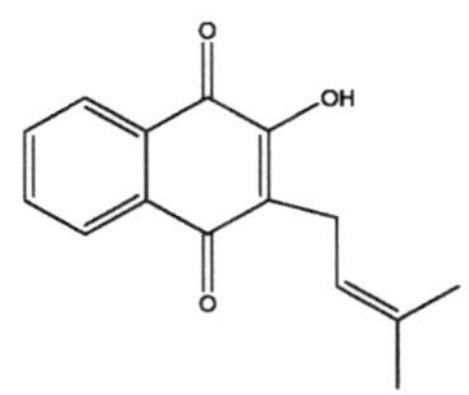	MTT assay HL-60 Western blotting flow cytometry D gel electrophoresis MTT assay MDA-MB231 4 T1 Immunofluorescence microscopy Experiments *in vivo* MHTBDE Huh7	IC_{50} μM 3.83 at 48 h Induced apoptosis in HL-60 strong alteration in cell proteome ERP57 is overexpressed in AML cells and is downregulated by shikonin IC_{50} μM 24 h 48 h 72 h 4.48 2.31 1.13 1.79 1.02 0.83 shikonin-mediated suppression of β-catenin signaling via increased levels of GSK-3β in MDA-MB-231 cells Shikonin inhibits lung metastasis and β-catenin signaling in NOD/SCID mice inoculated with MDA-MB-231 cells. IC_{50} M $5 X10^{-6}$	[21] [19] [22]

Quinona/ Natural origin (plants, animals, microorganisms)	Structure	Activity or model where it was tested	Results	Ref.
Juglone (5-hydroxy-1,4-naphthalenedione) *Juglans nigra* *Juglans mandshurica*	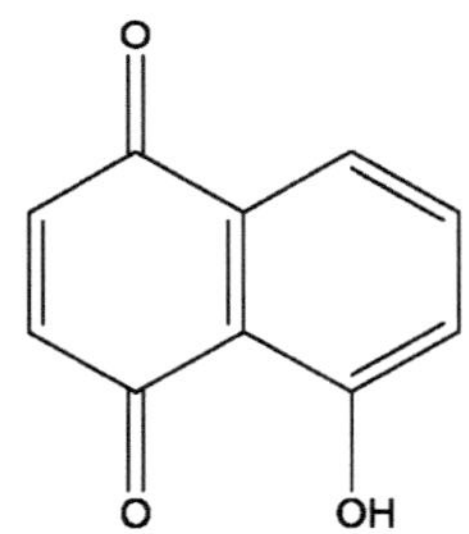	MTT assay F98 BGC-823 HCT-15 K562 HepG-2 WST-8 assay U87 SHG62 SHG66	% cell viability (50 μM) 20% (5 μM) 41% (0.5 μM) 59% IC_{50} μM 9.6 27.8 35.5 8.14 Cell viability % (10, 20, and 40 μM) 85, 60, 38 88, 62, 41 80, 55, 35	[23] [24] [25] [26]
Plumbagin *Nempenthes alata* Different species of *Plumbago* *Rumex dentatus, R. abyssinicus, R. usambarensis, R. bequaertii, R. ruwenzoriensis, R. crispus; Plumbago zeylanica,Myrsine Africana, Maesa lanceolata, Rapanea melanphloes, Aloe Saponaria* Several plants of the families: *Plumbaginaceae, Iridaceae, Drosophyllaceae, Droseraceae, Ebenceae* and *Nepenthaceae*	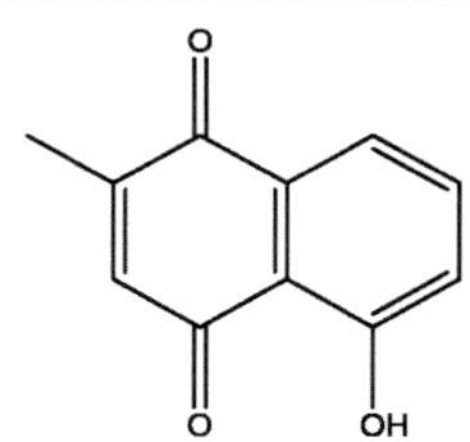	MTT assay MCF7 SK-OV-3 In mice bearing MCF7 cell xenografts MCF-10A MDA-MB231 MCF-7 Single cell gel electrophoresis assay. Clonogenic assay, Migration assay Western blot analysis NR assay A549 SPC212 DLD-1 Caco-2 MCF-7 HepG2 CRL2120. Annexin V-FITC binding assay MTT; Comet assay; PCR 786–0 cells	IC_{50} μM 3.5 13.1 Reduced tumor growth and weight without apparent side effects. Exerted its growth suppressive activity in MCF-7 by inducing apoptotic-related proteins. This compound is cytotoxic and caused cell membrane rupture in starting from 7.5 μM. Increase in the tail moment parameter with 7.5 μM. 3 1.5 3 Induced cytotoxicity in human breast cancer cells along with cell cycle arrest, DNA damage and cell death leading to apoptosis. Also found to suppress the telomerase activity in cancer cells accompanied by telomere attrition. IC_{50} μM 1.14	[27] [28] [29] [30]

Quinona/ Natural origin (plants, animals, microorganisms)	Structure	Activity or model where it was tested	Results	Ref.
			0.27 0.98 0.07 0.06 1.01 67.66 Reduced mRNA levels of MTOR and BCL2, and it did not affect the expression of CYP-encoding genes.	
Emodin (1,3,8-trihydroxy-6-methyl anthraquinone), *Rumex dentatus, R. abyssinicus, R. usambarensis, R. bequaertii, R. ruwenzoriensis, R. crispus; Plumbago zeylanica,Myrsine Africana, Maesa lanceolata, Rapanea melanphloes, Aloe saponaria* *Rheum palmatum* *Rhamnus sphaerosperma*	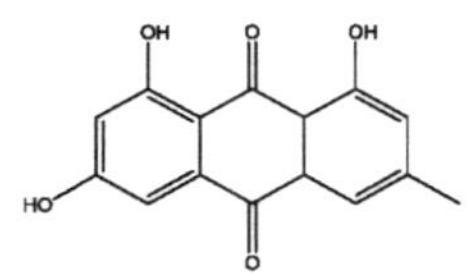	NR assay. A549 SPC212 DLD-1 Caco-2 MCF-7 HepG2 CRL2120. Flow cytometric assay Combination of IDH2 knockdown and emodin treatment on cell cycle disturbance. Cytomorphological Viability HaCaT SiHa C33A HSC-3 Annexin-V Cell Caspase-3 Activity Emodin using 12.5–50 μg/mL Western Blot DNA Damage Analysis	IC_{50} μM 66.3 99.31 77.28 73.63 37.57 71.7 >148.15 Suppression of IDH2 activity results in perturbation of the cellular redox balance and, ultimately, exacerbate emodin-induced apoptotic cell death in B16F10 cells. This result suggests that the combination of IDH2 downregulation and emodin treatment has negative effects on cancer cell growth Showed higher cytotoxic effects Induced apoptosis and necrosis independent of the caspase-3 activation pathway decreased the activation of AKT in all tumor cells, induction of reversible damage (DNA). Changed the Levels of BAX and BCL-2 Inhibited AKT.	[29] [31] [32]

Quinona/ Natural origin (plants, animals, microorganisms)	Structure	Activity or model where it was tested	Results	Ref.
Aloe-emodin (AE) *Rheum palmatum* and *Aloe vera*	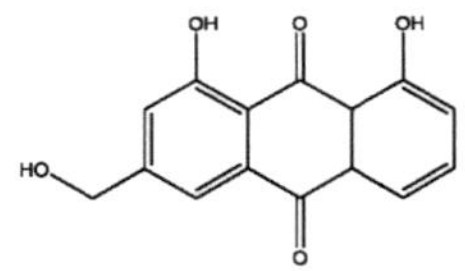	Formation of AE-derived glutathione conjugate (AE-GSH) and incubations containing AE and GSH, along with 3′-phosphoadenosine-5′-phosphosulfate (PAPS). The apoptotic induction by inverted phase contrast and fluorescence microscopes were used to evaluate apoptotic induction. Flow cytometry was used to determine the effects of aloe emodin on ΛΨm and cell cycle phase distribution.	AE undergoes sulfation, and the resulting AE-derived sulfate is chemically reactive to thiols. The phase II metabolism of AE may be a factor responsible for AE-induced cytotoxicity. This compound inhibits cancer cell growth MIAPaCa-2 and PANC-1 cell lines mediated by both ways, cell cycle arrest and loss of mitochondrial membrane potential.	[33] [34]
Fistulaquinones A (1). Fistulaquinones B(2). Fistulaquinones C (3). Isorhodoptilometrin-1-methyl ether (4). 7-hydroxyemodin-68-methyl ether (5). Sterequinone A (6). *Cassia fstula*	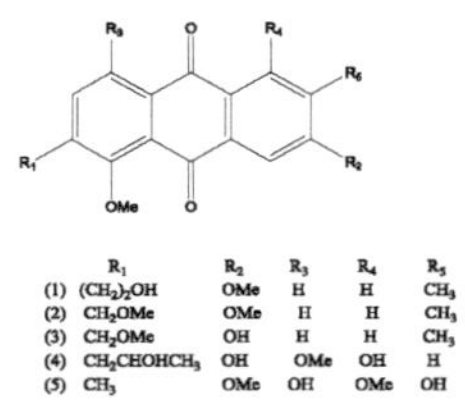 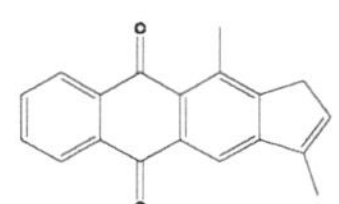 (6)	MTT-assay	IC_{50} μM NB4 A549 SHSYSY (1) 6.2 > 10 8.4 (2) >10 5.5 > 10 (3) 9 > 10 > 10 (4) 2.8 4.3 3.6 (5) >10 > 10 > 10 (6) 8.8 7.4 > 10 PC3 MCF7 (1) >10 > 10 (2) >10 > 10 (3) 7.2 > 10 (4) 4.2 5 (5) 9.4 > 10 (6) >10 5.5	[35]

Quinona/ Natural origin (plants, animals, microorganisms)	Structure	Activity or model where it was tested	Results	Ref.
Cyperaquinone (1) Hydroxycyperaquinone (2) Dihydrocyperaquinone (3) Tetrahydrocyperaquinone(4) Scabequinone (5) *Cyperus spp.*	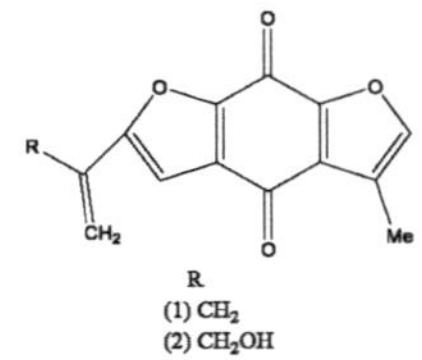 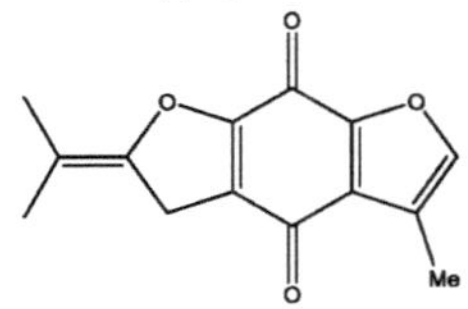 (3) 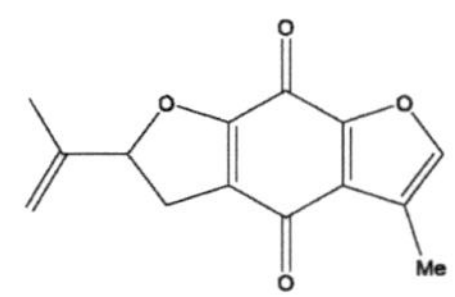 (4) 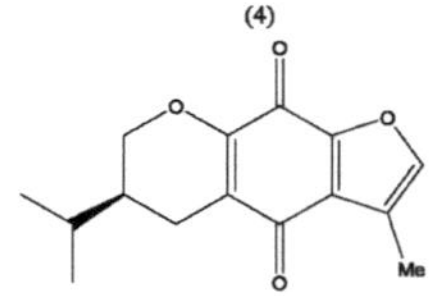 (5)	MTT assay Annexin V/7-AAD	IC_{50} μM A549 AGS MRC-5 (1) 11.3 3.0 8.7 (2) 3.0 1.7 1.7 (3) 45.3 1.8 > 50 (4) >50 > 50 > 50 (5) 46.6 27.4 28.7 None of the five compounds exert an effect upon caspase-9 activity nor caspase-3.	[36]
Cleistopholine *Enicosanthellum pulchrum*	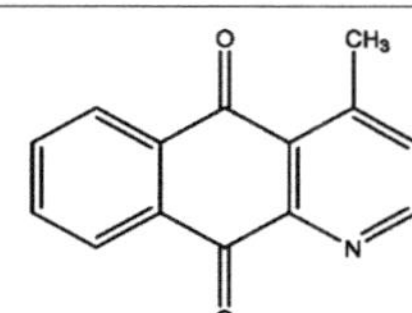	MTT assay CAOV-3 SKOV-3 Assessment of apoptosis morphology using acridine orange 86 (AO)/propidium iodide (PI) double staining Annexin-V-FITC.	IC_{50} μM 61.4 67.3 CAOV-3 cells showed morphological changes, evidenced by cell membrane blebbing, chromatin compression and formation of apoptotic bodies.	[37]

Quinona/ Natural origin (plants, animals, microorganisms)	Structure	Activity or model where it was tested	Results	Ref.
		Caspase 3, 8 and 9 Real-time PCR Western blot	Stimulated caspases 3 and 9 upregulated the mRNA expression levels of Bax, caspase 3 and caspase 9	
Primin *Eugenia hiemalis*	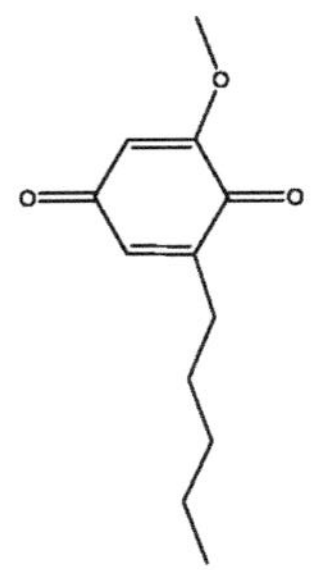	MTT assay K562 Jurkat MM.1S	IC_{50} μM 24 h 48 h 72 h 7.54 4.93 2.65 4.16 1.50 0.55 5.31 5.11 1.36 Apoptosis appears to be related to a decreased Bcl-2 expression and increased Bax expression	[38]
marcanine G *Goniothalamus marcanii*	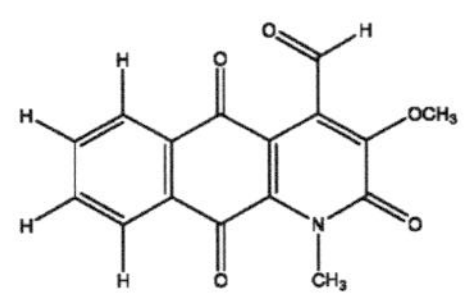	SRB assay A549 MCF-7 MRCS	IC_{50} μM 14.87 15.18 15.45	[39]
2,7-Dihydroxy-3-methylanthraquinone (DDMN) *Hedyotis difusa*	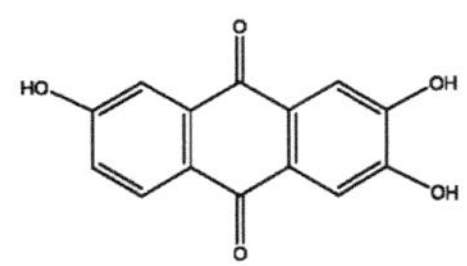	MTT assay SGC-7901 Flow cytometry assay. Xenograft assay	IC_{50} μM 20.92 Induces death by apoptosis. Tumor growth on nude mice could be significantly inhibited during the 20 days period (40 mg / kg / d)	[40]

Quinona/ Natural origin (plants, animals, microorganisms)	Structure	Activity or model where it was tested	Results	Ref.
2-methoxy-1,4-naphthoquinone *Impatiens glandulifera*	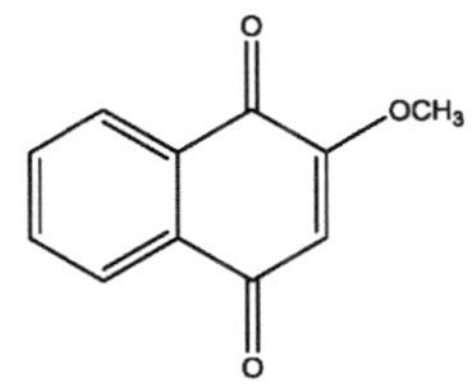	MTT assay A549 SKMEL-28 V373	IC_{50} μM. 3 2 3	[41]
5-methoxy-1,4-naphthoquinone (1). 5,8-dihydroxy-1,4-naphthoquinone (2). 2-hydroxy-1,4-naphthoquinone (3). 2,5-dihydroxy-1,4-naphthoquinone (4). 3,5-dihydroxy-1,4-naphthoquinone (5). 3-methoxy juglone (6). 2-methoxy juglone (7). 3-ethoxy juglone (8). 2-ethoxy juglone (9). Engelharquinone (10). *Juglans mandshurica*	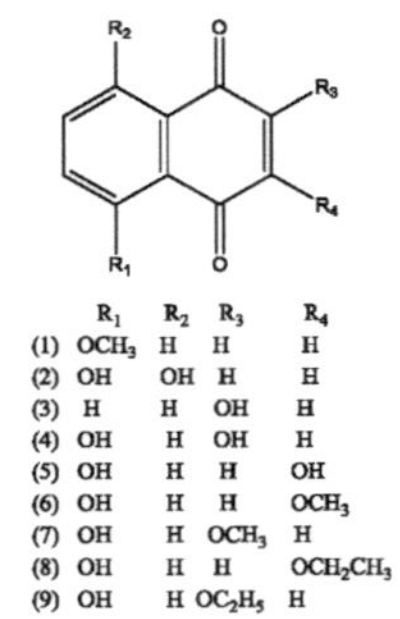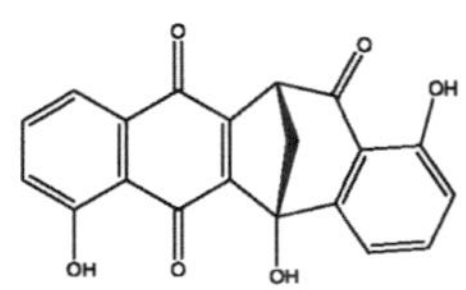(10)	MTT assay HepG-2	IC_{50} μM (1) 68.72 (2) 16.11 (3) 18.83 (4) 15.37 (5) 7.33 (6) 43.54 (7) 22.38 (8) 30.42 (9) 32.51 (10) 34.80	[25]

Quinona/ Natural origin (plants, animals, microorganisms)	Structure	Activity or model where it was tested	Results	Ref.
2-hydroxy-1,4-naphthoquinone (1). 2,5-dihydroxy-1,4-naphthoquinone (2). 5,8-dihydroxy-1,4-naphthoquinone (3). 3,5-dihydroxy-1,4-naphthoquinone (4). *Juglans mandshurica*	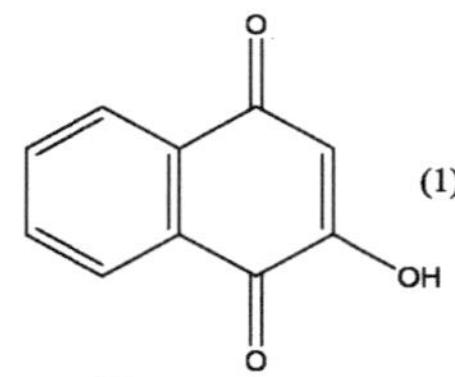 (1) 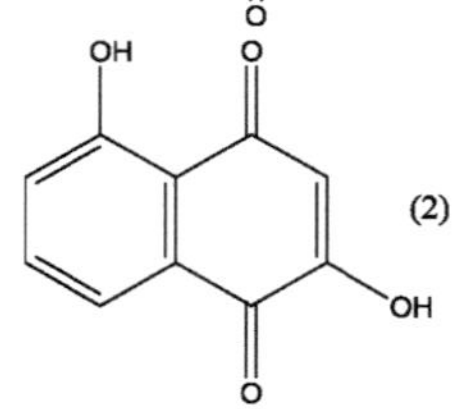 (2) 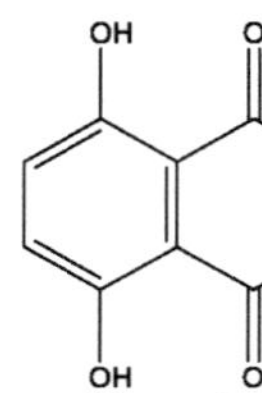 (3) 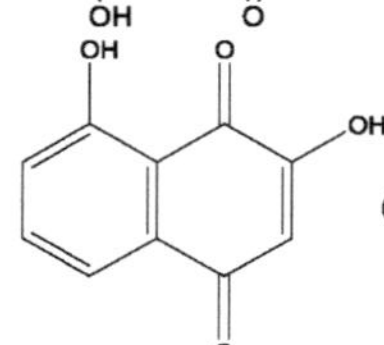 (4)	MTT Assay	IC_{50} μM BGC-823 HCT-15 K562 (1) —— 37.4 —— (2) 33.8 97.9 39.7 (3) 28.2 —— —— (4) 19.0 —— ——	[24]

Quinona/ Natural origin (plants, animals, microorganisms)	Structure	Activity or model where it was tested	Results	Ref.
Knipholone *Kniphofia foliosa* Hochst	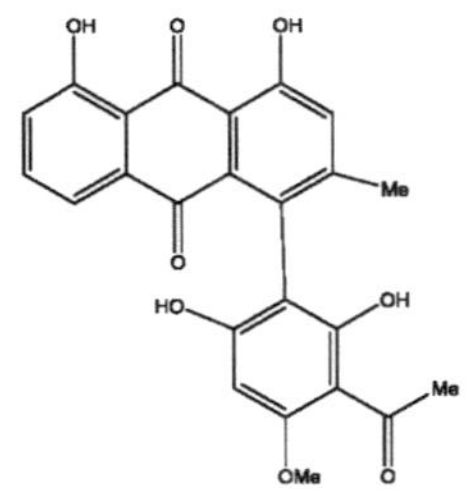 	MTT assay Jurkat HEK29 SH-SY5Y	Cell viability % 62–95% at 50 μM	[42]
β,β-dimethylacrylshikonin *Lithospermum erythrorhizon*	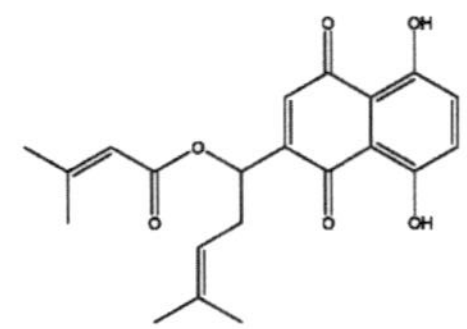 	MTT assay MDA MB231 4 T1	IC_{50} μM 24 h 48 h 72 h 18.7 11.6 4.30 14.7 7.88 4.13	[19]
Mansonone-G (MG). *Mansonia gagei*	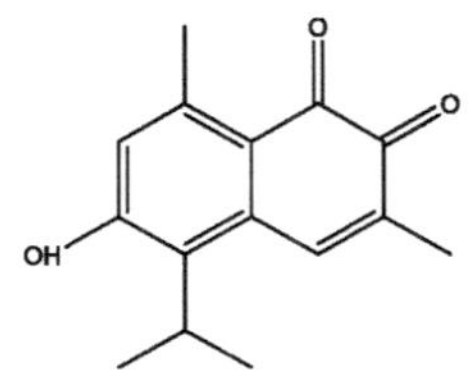 	SRB assay MCF HeLa HCT-116 HepG2	IC_{50} μM 23 18.8 63.4 49.4	[43]
2-acetyl-7-methoxynaphtho[2,3-b]furan-4,9-quinone *Milletia versicolor*	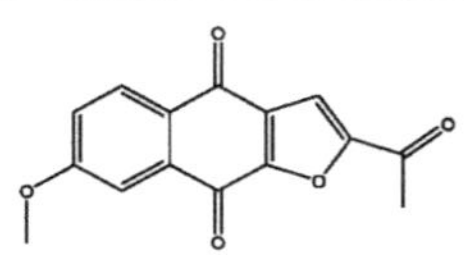 	The resazurin reduction assay CCRF-CEM CEM/ADR5000 MDA-MB231 MDAB231/*BCRP* HCT116 *(p53+/+)* HCT116*(p53−/−)* U87MG U87MG.*ΔEGFR* HepG2 AML12	IC_{50} μg/mL 0.16 0.28 0.58 0.89 0.27 0.61 0.27 0.26 0.22 >40	[44]

Quinona/ Natural origin (plants, animals, microorganisms)	Structure	Activity or model where it was tested	Results	Ref.
Nordamnacanthal (NDAM) *Morinda citrifolia* L	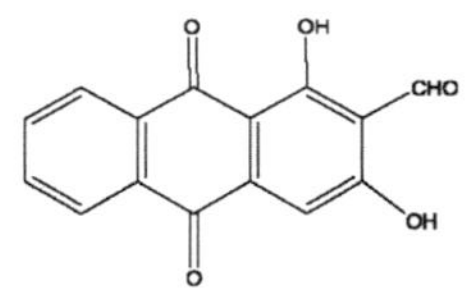	MTT assay TBE assay In vivo study of the antitumor effect of NDAM using 4 T1-bearing BALB/C mice. Flow cytometry Immunophenotyping analysis of CD3, CD4 and CD8-stained splenocytes.	IC_{50} μg/mL MDA-MB-231 4TI MCF-7 12.5 12.5 11 1.2 10 8 NDAM reduced the 4 T1 tumor size and weight. Cease the tumor progression of 4 T1 cells *in vivo*. Induced apoptosis in MCF-7, MDA-MB231 and 4 T1 cells *in vitro* NDAM regulated several immune markers in tumor-bearing mice	[45]
Thymoquinone *Nigella sativa*	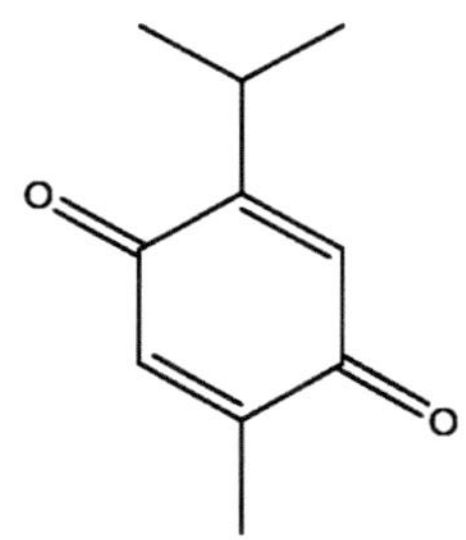	MTT assay EMT6/P MCF-7 T47D Vero-normal MRC-5 Neuro-2a Wound healing assay	IC_{50} μM. 393 55 85 45 IC_{50} μg/mL Non-covered plates 2.95 EVA capmat™ covered plates 1.72 IC_{50}, μM 24 h 48 h 20 40 Inhibitory effect on the migration of Neuro-2a cells was mediated through the suppression of MMP-2 and MMP-9 expression.	[46] [47] [48]

Quinona/ Natural origin (plants, animals, microorganisms)	Structure	Activity or model where it was tested	Results	Ref.
Deoxyshikonin (1). Isobutyrylshikonin (2). α-methylbutyrylshikonin (3). β- hydroxyisovalerylshikonin (4). 5,8-*O*-dimethyl isobutyrylshikonin (5). 5,8-*O*-dimethyl deoxyshikonin (6). *Onosma visianii*	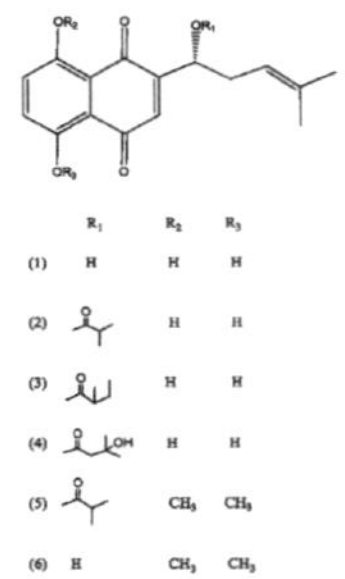	MTT cell viability assay Cell cycle analysis	IC_{50} μg/mL 72 h MDA-MB-231 HCT-116 (1) 119 98 (2) 425 202 (3) 86 15 (4) 205 301 (5) 412 128 (6) 392 485 All compounds induce cell cycle arrest in tumor cell lines.	[20]
1-hydroxy-6,8-dimethoxy-3-methylanthracene-9, 10-dione (1). 8-hydroxy1,3-dimethoxy-6-methylanthraquinone (2). xanthopurpurin (3). 2-methyl-1,3,6-trihydroxy-9,10-anthraquinone (4). *Osmunda japonica*	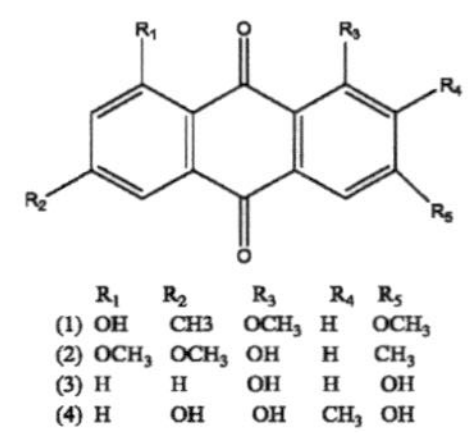	MTT assay Hela HepG2 A549	IC_{50} μg/mL Weak activity	[49]
Physcion *Osmunda japonica* *Rhamnus sphaerosperma*	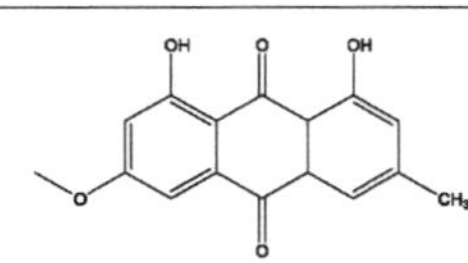	MTT assay Hela, HepG2 A549 Cytomorphological Viability HaCaT SiHa C33A HSC-3 Annexin-V Cell Caspase-3 Activity Physcion using 12.5–50 μg/mL Western Blot DNA Damage Analysis	IC_{50} μg/mL Weak activity Showed higher cytotoxic effects Induced apoptosis and necrosis independent of the caspase-3 activation pathway decreased the activation of AKT in all tumor cells, induction of reversible damage (DNA). Changed the Levels of BAX and BCL-2 Inhibited AKT.	[49] [32]

Quinona/ Natural origin (plants, animals, microorganisms)	Structure	Activity or model where it was tested	Results	Ref.
2,5-dihydroxy-3-heptyl-2,5-cyclohexadiene-1,4-dione (1) 2,5-dihydroxy-3-tridecyl-2,5-cyclohexadiene-1,4-dione or rapanone (2). 2,5-dihydroxy-3-pentadecyl-2,5-cyclohexadiene-1,4-dione (3). Adisiaquinone B (4) *Rumex dentatus, R. abyssinicus, R. usambarensis, R. bequaertii, R. ruwenzoriensis, R. crispus; Plumbago zeylanica,Myrsine Africana, Maesa lanceolata, Rapanea melanphloes, Aloe saponaria*	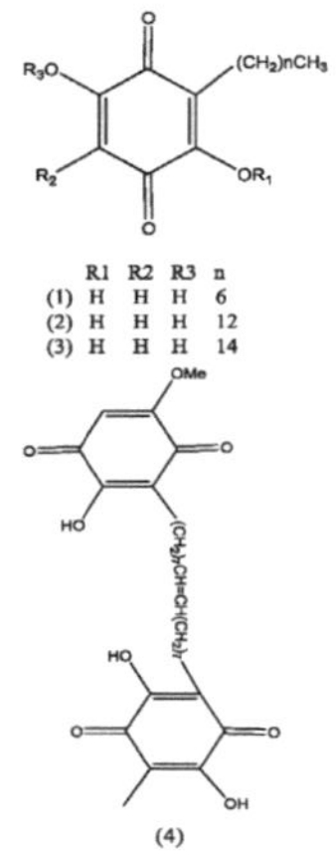	Neutral red uptake (NR) assay A549, SPC212, DLD-1, Caco-2, MCF-7, HepG2, CRL2120.	IC_{50} µM (1) 8.05 to 117.27 (2) 2.27 to 46.62 (3) 8.39 to 48.35 (4) 3.14 to 114.17	[29]
Alizarin (1). Xanthopurpurin (2) lucidin-ω-methyl ether (3). *Rubia philippinensis*	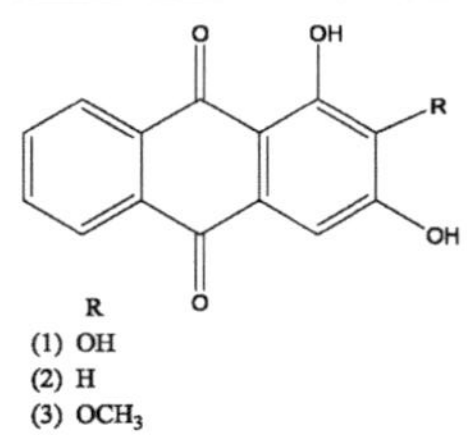	MTT assay	IC_{50} µM SK-MEL B16F10 MCF-7 MDA-MB231 (1) 53.08 98.79 49.17 48.64 (2) 21.35 23.71 15.75 14.65 (3) 42.79 10.51 8.59 7.95	[50]

Quinona/ Natural origin (plants, animals, microorganisms)	Structure	Activity or model where it was tested	Results	Ref.
Rhinacanthin-C *Rhinacanthus nasutus*	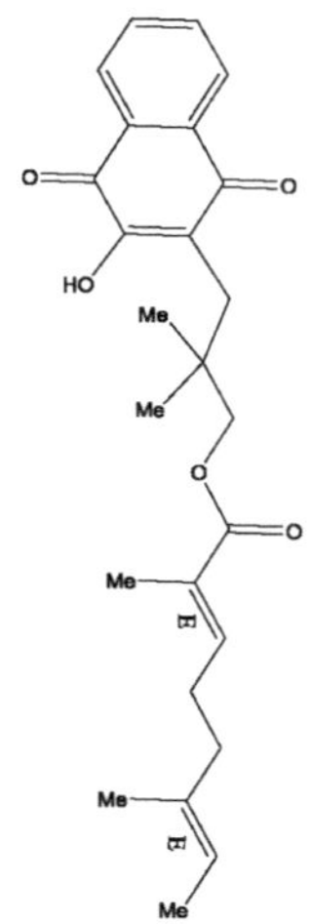	Sulforhodamide B assay KKU-M156 Vero Wound migration assay Chamber migration assay Chamber invasion assay. Gelatin zymography and uPA assay. Western blot analysis	IC_{50} μM 1.50 2.37 Inhibits the migration and invasion by decreasing MMP-2, uPA, FAK and MAPK pathways	[51]
Rhinacanthin S *Rhinacanthus nasutus*	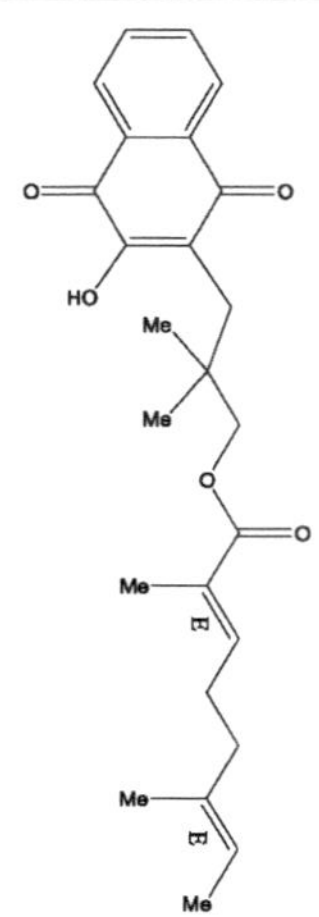	RM assay KB MCF-7 NCI-H148	IC_{50} μM 11.66 20 15.86	[52]

Quinona/ Natural origin (plants, animals, microorganisms)	Structure	Activity or model where it was tested	Results	Ref.
β-Lapachone *Tabebuia avellanedae* *Tabebula impetiginosa*	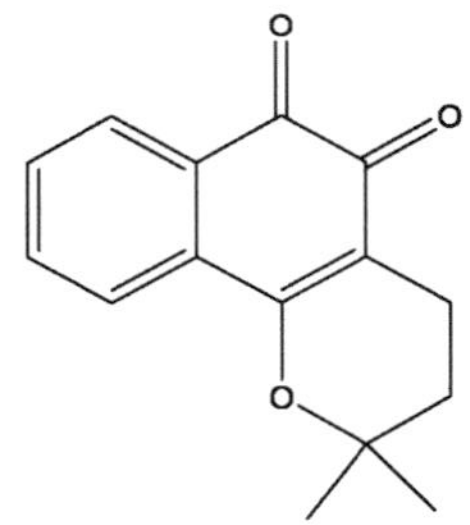	ABR assay HSC3, SCC4, SCC9, SCC15 SCC25 HepG2 HL-60 K562 AGP01 ACP-02 ACP-3 HT-29 HCT-116 FITC Annexin V Apoptosis qPCR array Westerm blot analysis	IC_{50} μM 1.02 16.22 0.16 0.06 2.78 0.99 0.09 1.35 20.33 48.94 15.49 25.03 5.62 Induce cell cycle arrest at G2/M phase and promote caspase- and ROS-mediated apoptosis In total, 44 genes were investigated in HSC3 cells treated with β-lapachone the pro-apoptotic genes BAX. Induced apoptotic cell death by NQO1-mediated ROS in a dose-dependent manner on MDA-MB-231 cells overexpressing NQO1 (231-NQO1+/+) MDA-MB-231 cells lacking NQO1 (231-NQO1−/−).	[53] [54]
3-diethylamino-5-methoxy-1, 2-benzoquinone (1) 3-ethylamino-5-methoxy-1, 2-benzoquinone (2) *Uncaria rhynchophylla*	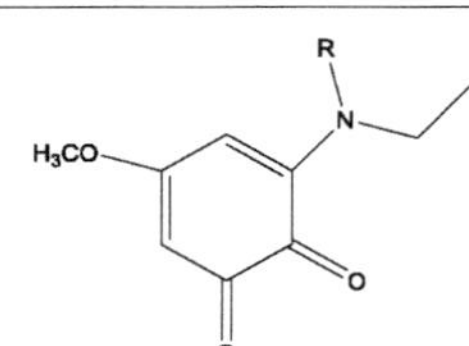 R (1) CH_2CH_3 (2) H	MTT assay	IC_{50} μM A549 HepG2 A2780 (1) 50.2 97.2 84.6 (2) 94.8 > 100.0 98.8	[55]

Quinona/ Natural origin (plants, animals, microorganisms)	Structure	Activity or model where it was tested	Results	Ref.
Ventilanone A (1) Ventilanone B (2) Ventilanone C (3) Ventilanone D (4) Ventilanone E (5) *Ventilago harmandiana*	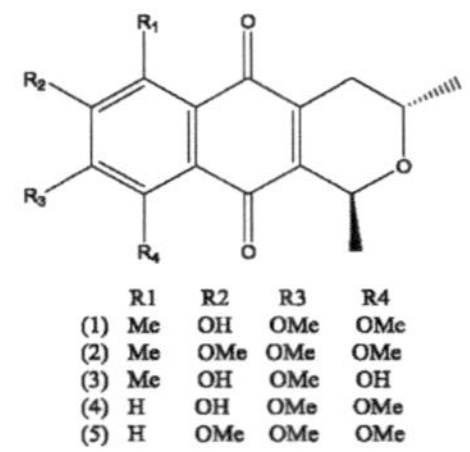	SRB assay	ED_{50} μM P-388 KB Col-2 (1) 9.33 > 50 > 50 (2) >20 > 50 > 50 (3) 13.82 > 50 > 50 (4) >20 > 50 > 50 (5) >20 37.31 38.81	[56]
Marine sponge				
Smenospongiarine (1) Smenospongorine (2) Smenospongimine (3) *Dactylospongia elegans*	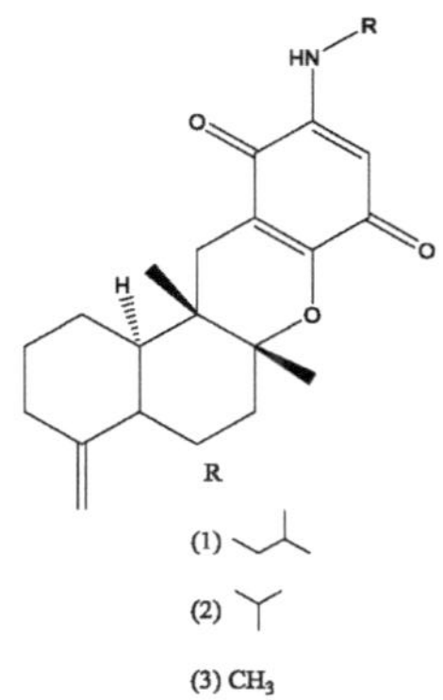	CCK-8 method DU145 SW1990 Huh7 PANC-1	IC_{50} μM ranging from 2.33 to 37.85	[57]

Quinona/ Natural origin (plants, animals, microorganisms)	Structure	Activity or model where it was tested	Results	Ref.
Ilimaquinone (1) 5-epi-ilimaquinone (2) smenospongine (3) smenospongorine (4) smenospongiarine (5) smenospongidine (6) *Dactylospongia elegans*	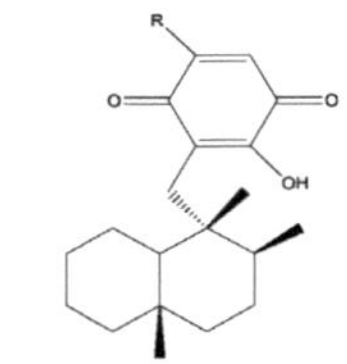 R (1) OMe (2) OMe C5 epimer (3) NH_2 (4) $NHCH_2CH(CH_3)_2$ (5) $NHCH_2CH_2CH(CH_3)_2$ (6) $NHCH_2CH_2Ph$	Commertial Kit C	CC_{50} μM U251MG Panc-1 (1) 19.3 20.4 (2) 19.4 16.2 (3) 2.4 —— (4) 19.4 22.6 (5) 4.5 15.1 (6) 4.0 12.6	[58]
(+)-19-methylaminoavarone(1) (−)-20-methoxyneoavarone(2) (+)-20-methoxyavarone(3) (+)-17,20-dimethoxyavarone(4) (+)-19-methoxyavaron(5) (−)-20-phenethylaminoavarone (6) (−)-20-methylaminoavarone(7) The different spices of *Dysidea sp.*	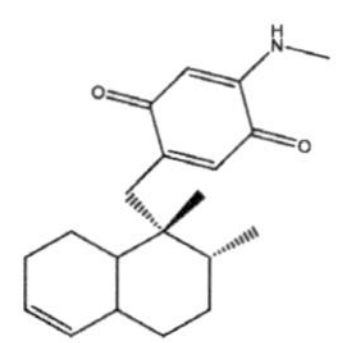 (1) 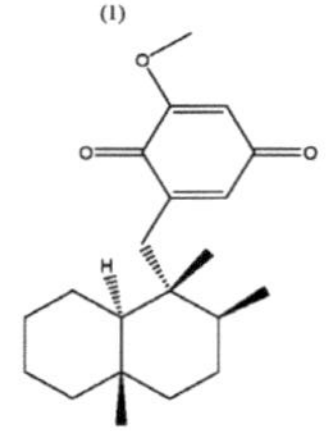 (2) 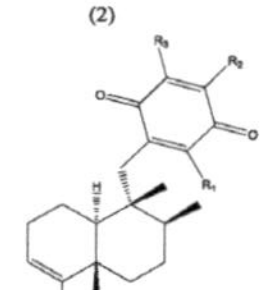 R1 R2 R3 3) H H OCH_3 4) OCH_3 H OCH_3 5) H OCH_3 H 6) H H $NHCH_2CH_2Ph$ 7) H H $NHCH_3$	MTT assay A549 Hela HCT-116 Jukat K562 BEL-7402	IC_{50} μM Compound 2 showed the best cytotoxic activity with IC_{50} values ranging from 0.93 to 4.61 mM	[59]

Quinona/ Natural origin (plants, animals, microorganisms)	Structure	Activity or model where it was tested	Results	Ref.
Langcoquinones D The different genera *Dysidea*, *Spongia*, and *Dactylospongia*	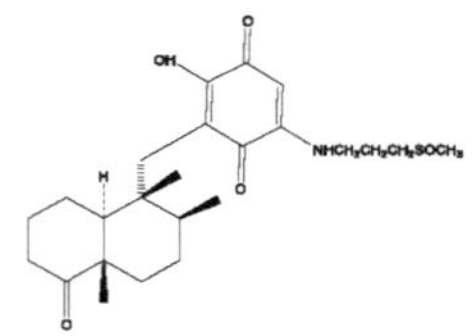	WST-8 cell counting kit solution A549 MCF-7 HeLa	IC_{50} µM 8.9 5.9 8.6	[60]
Fugus				
Antrocinnamone (1) Quinone Q3 (2) Antrocamol LT3 (3) Antroquinonol (4) Antroquinonol B (5) *Antrodia cinnamomea*	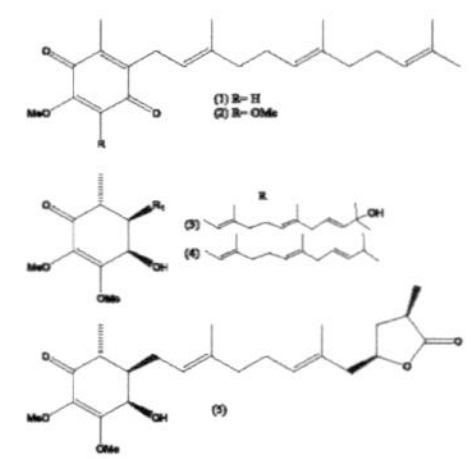	The cell counting kit-8 assay	IC_{50} µM MDCK A549 HepG2 PC3 (1) >100 0.382 > 100 0.014 (2) >100 4.16 > 100 0.060 (3) >100 0.008 0.106 0.001 (4) 10.53 0.421 0.044 0.073 (5) >100 6.032 21.37 1.031	[61]
6,6′-oxybis(1,3,8-trihydroxy-2-((S)-1-methoxyhexyl)anthracene-9,10- dione) (1). 6,6′-oxybis(1,3,8-trihydroxy-2-((S)-1-hydroxyhexyl) anthracene9,10-dione) (2). 1′-O-methylaverantin (3). Averantin (4) Averythrin (5) *Aspergillus versicolor*	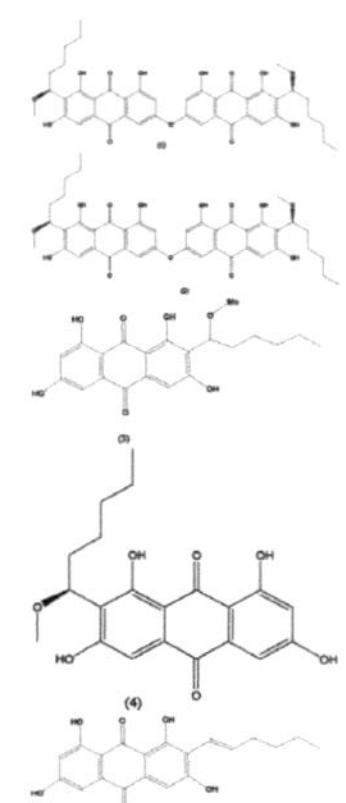	MTT assay SK-OV-3 SK-MEL-2 CNS XF498 HCT-15	IC_{50} µg/mL values ranging from 11.25–17.36	[62]

Quinona/ Natural origin (plants, animals, microorganisms)	Structure	Activity or model where it was tested	Results	Ref.
Cochlioquinones G (1) Cochlioquinones H (2) Cochlioquinone C (3) Cochlioquinone E (4) Cochlioquinone B (5) Cochlioquinone D (6) *Bipolaris sorokiniana*	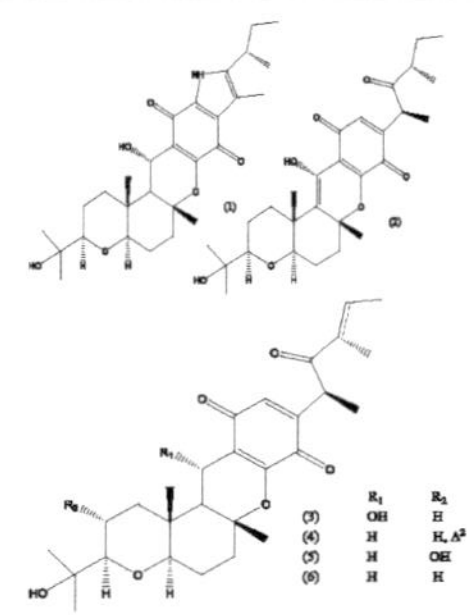	SRB assay.	IC_{50} μM SF-268 HepG-2 MCF7 (1, 2, 3, 4, 5) IC_{50s} < 10 μM (6) 1.5 2.4 1.2	[63]
Purpurogemutantin (1). Macrophorin A (2). 4′-oxomacrophorin (3). *Gliomastix sp.* ZSDS1-F7	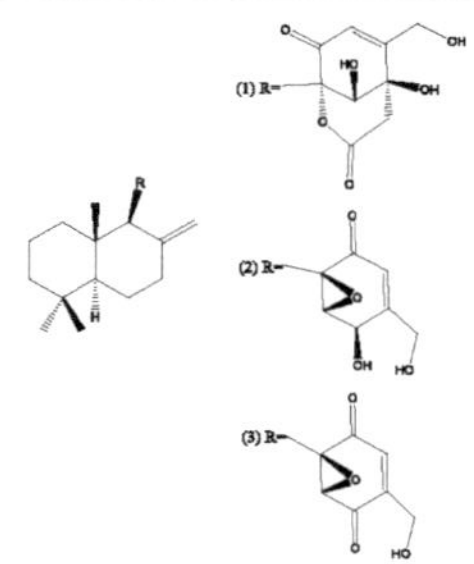	CCK-8 method K562, MCF-7, Hela, DU145, U937, H1975, SGC-7901, A549, MOLT-4 and HL60 cell lines	IC_{50} values ranging from 0.19 to 35.4 μM.	[64]
Ophioparmin (1). 4-methoxyhaemoventosins (2). 4-hydroxyhaemoventosin (3). *Ophioparma ventosa* lichen	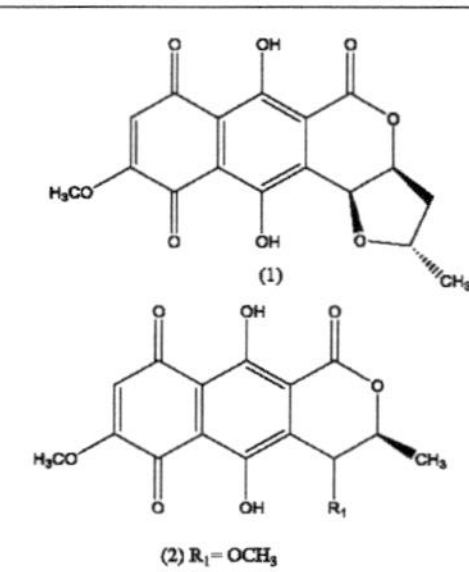	MTT assay B16 HaCaT	IC_{50} μg/mL >10	[65]

Quinona/ Natural origin (plants, animals, microorganisms)	Structure	Activity or model where it was tested	Results	Ref.
Peniquinone A (1) Peniquinone B (2) *Penicillium* sp. L129	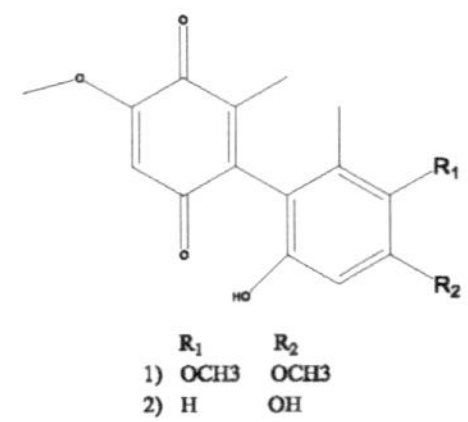	MTT assay	IC_{50} μM MCF-7 A549 U87 PC3 (1) 12.39 > 40 9.01 14.59 (2) 25.0 > 40 13.45 19.93	[66]
Altersolanol A *Phomopsis* sp. (PM0409092)	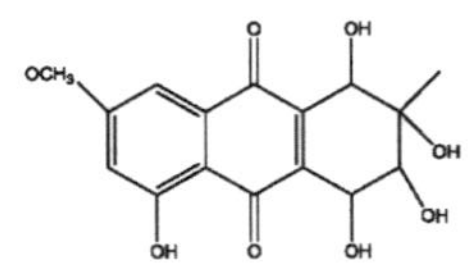	Monolayer assay propidium iodide (PI)	IC_{50} μg/mL	[67]
		BXF 1218 L	0.001	
		BXF T24	0.001	
		CNXF 498NL	0.001	
		CNXF SF268	0.001	
		CXF HCT116	0.287	
		CXF HT29	0.001	
		GXF 251 L	0.052	
		LXF 1121 L	0.004	
		LXF 289 L	0.027	
		LXF 526 L	0.001	
		LXF 529 L	0.004	
		LXF 629 L	0.001	
		LXF H460	0.412	
		MAXF 401NL	0.01	
		MAXF MCF7	0.001	
		MEXF 394NL	0.001	
		MEXF 462NL	0.034	
		MEXF 514 L	0.001	
		MEXF 520 L	0.001	
		OVXF 1619 L	0.001	
		OVXF 899 L	0.006	
		OVXFOVCAR	0.013	
		PAXF 1657 L	0.049	
		PAXF PANC1	0.001	
		PRXF 22RV1	0.012	

Quinona/ Natural origin (plants, animals, microorganisms)	Structure	Activity or model where it was tested	Results	Ref.
		PRXF DU145 PRXF LNCAP PRXF PC3M PXF 1752 L RXF 1781 L RXF 393NL RXF 486 L RXF 944 L UXF 1138 L	0.001 0.01 0.01 0.05 0.075 0.035 0.095 0.001 0.061	
2,6-dimethoxy-1,4-benzoquinone *Saccharomyces cerevisiae*	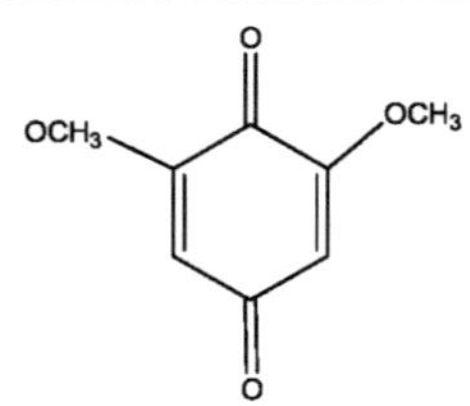	CV assay MDA-MB-468 MDA-MB-231 BT-20 MCF-7 23132/87, ASPC-1, BxPC-3, HT-29, HRT-18b	IC_{50} mg/mL 3.8 5.5 13.3 19.3 7.9 4.0 4.4 10.9 15.8	[68]
Auxarthrol D (1) Auxarthrol F (2) *Sporendonema casei* HDN16–802	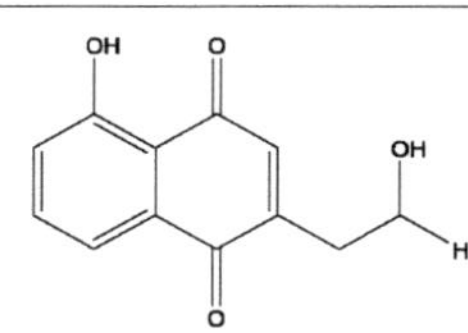 (1) 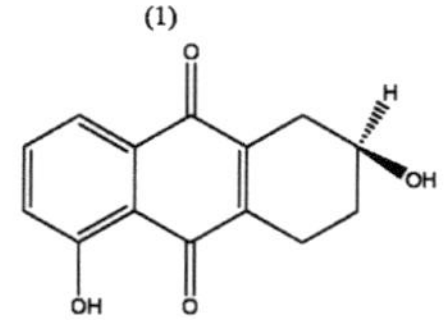 (2)	MTT assay HL-60; Hela; HCT-116; MGC-803; HO8910; MDA-MB-231; SH-SY5Y; PC-3; BEL-7402; K562; L-02.	IC_{50} μM values ranging from 4.5 to 22.9	[69]

Quinona/ Natural origin (plants, animals, microorganisms)	Structure	Activity or model where it was tested	Results	Ref.
5- hydroxy-2-(2-hydroxypropyl)naphthalene-1,4-dione (1). (S)-2,5-dihydroxy-2-methyl1,2,3,4-tetrahydroanthracene-9,10-dione (2). *Micromonospora* sp. NEAU-gq13	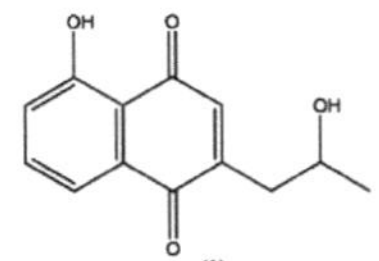 (1) 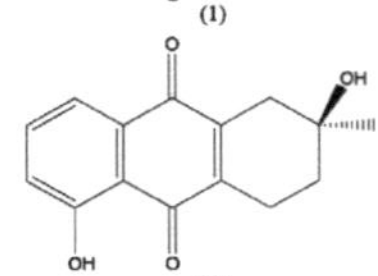 (2)	CCK-8 colorimetric method	IC50 μg/ml HepG2 SF-268 ACHN (1) 1.01 3.04 10.08 (2) 12.98 5.66 11.43	[70]
Bacteria				
2-amino-6-hydroxy-[1,4]-benzoquinone *Geobacillus* sp. E263	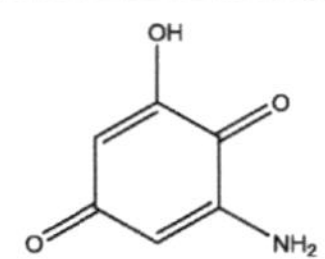	Detection of apoptosis for fluorescence assay	The percentage of apoptotic cancer cells (MGC-803, HGC-27, MDA-MB-231,MDA-MB-435) at 10 or 100 μM was significantly increased	[71]
Napyradiomycin A3 (1) Napyradiomycin B7a (2) Napyradiomycin B7b (3) Napyradiomycin SC (4) *Streptomyces* sp. strain CA-271078	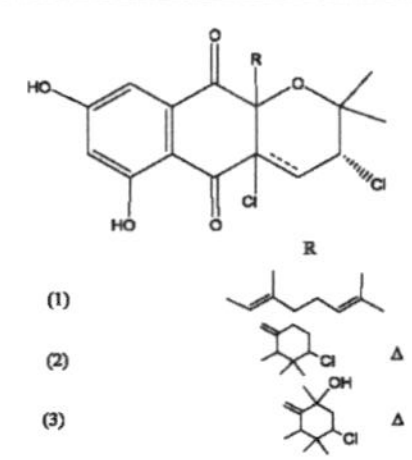 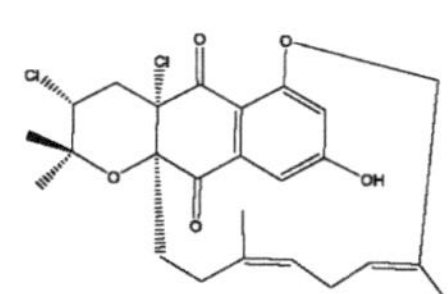 (4)	MTT assay HepG2	IC_{50} μM Values >50	[72]

Quinona/ Natural origin (plants, animals, microorganisms)	Structure	Activity or model where it was tested	Results	Ref.
Abenquine A (1) Abenquine B1 (2) Abenquine B2 (3) Abenquine C (4) Abenquine D (5) *Streptomyces sp.* strain DB634	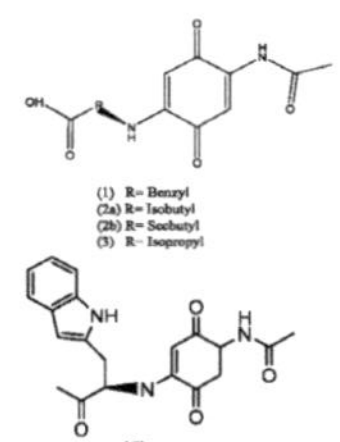 (5)	SRB assay 518A2 A2780 HT29 MCF7 A549 FaDu NIH 3 T3	EC_{50} μM The compounds on the the 7 cell lines showed values >30	[73]
Medermycin *Streptomyces* sp. SS17A	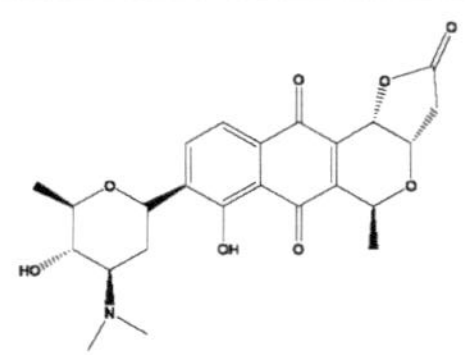	MTT assay PC3 HCT-116	IC_{50} μM 0.02 0.04	[74]
Sharkquinone *Streptomyces sp.* EGY1	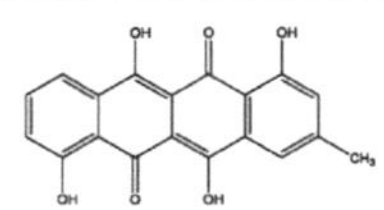	CAF AGS	IC_{50} μM 7.3	[75]

Celular lines: 23132/87, BGC-823, GXF 251 L and SGC-7901 human gastric carcinoma cells; ASPC-1, BxPC-3, PAXF 1657 L and PAXF PANC1 adenocarcinoma of the pancreas; DLD-1, Caco-2, HCT-15, CXF HCT116, CXF HT29, HCT116, HRT-18 and HT29 colon adenocarcinoma cells; PRXF 22RV1, PRXF DU145, PRXF LNCAP, PRXF PC3M and DU-145 human prostate carcinoma; MCF-7, MDA-MB 231, MDA-MB-468, BT-20 and BT-474 Human breast carcinoma cell line; A2780, OVXF 1619 L, OVXF 899 L, OVXF OVCAR3,CAOV-3 and SKOV-3 Human ovarian cancer cells; A549, H-1299, LXF 1121 L, LXF 289 L, LXF 526 L, LXF 529 L, LXF 629 L, LXF H460, NCI-H187, NCI-H1437, NCIH1655, NCI-H358, NCI-H460 and NSCLC cancer lung cells; CNXF 498NL and CNXF SF268 cancer of Central nervous System cells; RXF 1781 L, RXF 393NL, RXF 486 L, RXF 944 L and ACHN human renal cancer; BXF 1218 L and BXF T24 cancer Bladder cells; XF498 and SF-268 human central nervous system cancer; MEXF 394NL, MEXF 462NL, MEXF 514 L, MEXF 520 L, SK-MEL-5, 518A2, B16F10, C33A, HSC3, SCC4, SCC9, SCC15, SCC25 melanoma cells; CRL2120 and SK-MEL-2 human skin cancer; BEL-7402 and HepG2; SiHa, KB3.1 and HeLa human cervical adenocarcinoma cells; Jurkat lymphoblastic, HL-60 and K562 leukemia cells; MIAPaCa-2 and PANC-1 human pancreatic adenocarcinoma cancer; UXF 1138 L cancer Uterus cells; OC3-IV2 Human oral cancer; PXF 1752 L pleuramesothelioma; SPC212 human mesotelioma cell; SYF mouse embryonic fibroblast deficient in C-Src; U251MG human glioblastoma; FaDu hypopharyngeal carcinoma; KKU-M156 human cholangiocarcinoma cells. WI38 human normal lung, HaCaT immortalized human keratinocytes, nontumorigenic cell line, Vero kidney of a normal monkey cell, L929 nonmalignant mouse fibroblasts and NIH 3 T3 nonmalignant mouse fibroblasts. Inhibitory concentration of 50% (IC_{50}); Effective concentration of 50% (EC_{50}); Growth inhibition of 50% (IG_{50}); Assay of 3-(4,5-dimethylthiazol-2-yl)-2,5-diphenyltetrazolium bromide (MMT); Microscopy on a hemocytometer using trypan blue dye exclusion method (MHTBDE); Cytotoxicity Assay for fluorescence (CAF); Neutral red uptake assay (NR); Sulforhodamine B assay (SRB); Trypan Blue Exclusion assay (TBE); Alamar Blue reduction assay (ABR); Resazurin microplate assay (RM); Crystal violet assay (CV).

Table 1.
Anticancer activity of quinones isolated from different organism.

Quinona/ Natural origin (plants, animals, microorganisms)	**Structure**	**Activity or model where it was tested**	**Results**	**Reference**
Plants				
Deoxyshikonin (***1***), isobutyrylshikonin (***2***), α-methylbutyrylshikonin (***3***), acetylshikonin (***4***), β-hydroxyisovalerylshikonin (***5***), 5,8-*O*-dimethyl isobutyrylshikonin (***6***) and 5,8-*O*-dimethyl deoxyshikonin (***7***). *Onosma visianii*	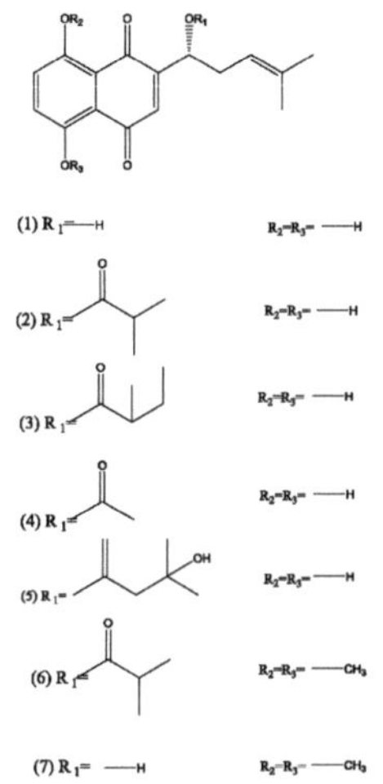	Micro-dilution antibacterial assay *B. megaterium* *E fecalis* *M. arborescens* *M. luteus* *S. epidermis* *C. Koseri* *H alvei* *P. proteolytica* *S. maltophilia* *Y. intermedia*	MIC 50 and 90 μg/mL For all compounds Range: 8–51/ 9–54.28 6–34/6–38 6–34/6–38 8–68/9–76 8–51/9–54 6–68/6–76 6–51/6–54 4–68/6–38 4–68/4–76 6–25/6–76	[20]
Thymoquinone *Nigella sativa*	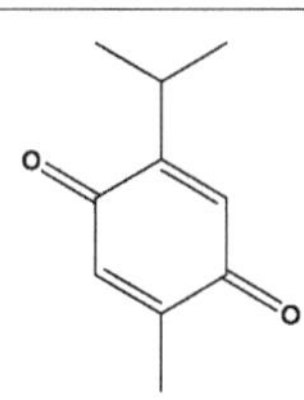	Broth microdilution volatilization method *Haemophilus influenzae* *Staphylococcus aureus* *Streptococcus pneumoniae*	MIC (Broth/ agar) μg/mL 8/8 16/16 16/32	[47]
1,4,6-Trihydroxy-8-isoheptanyl-9,10-anthraquinone (symploquinone A) (1) 1,4-Dihydroxy-6-methyl-8-isopropyl-9,10-anthraquinone (symploquinone C) (2) *Symplocos racemosa*	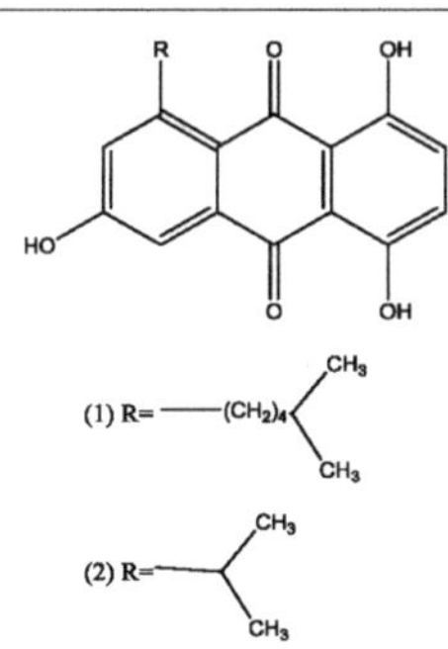	Microdilution assay *S. aureus* *P mirabilis*	MIC μg/mL (1) 160 (2) 83 (1) >160 (2) >160	[76]
Primin *Miconia willdenowii*	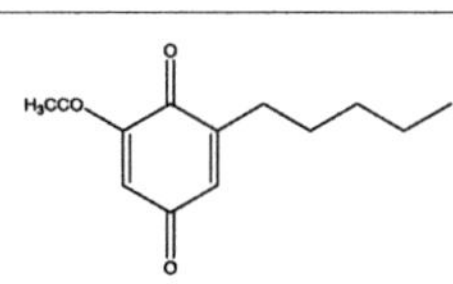	Mueller Hinton broth microdilution assay *C. albicans ATCC 10231* *C. krusei ATCC 6258* *C. tropicalis ATCC 750* *C. glabrata ATCC 90030* *C. parapsilosis ATCC 22019* *S. aureus (ATCC 6538)*	IC_{50} μM 72.08 36.04 72.08 72.08 72.08 8.94	[77]

Quinona/ Natural origin (plants, animals, microorganisms)	Structure	Activity or model where it was tested	Results	Reference
Lawsone *Lawsonia inermis*	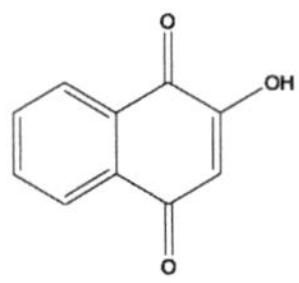	Microdilution assay *Saccharomyces cerevisiae*. Strain BY4741	MIC mM/L 229	[78]
Marine Sponge				
langcoquinones D *Dysidea*, *Spongia* and *Dactylospongia*	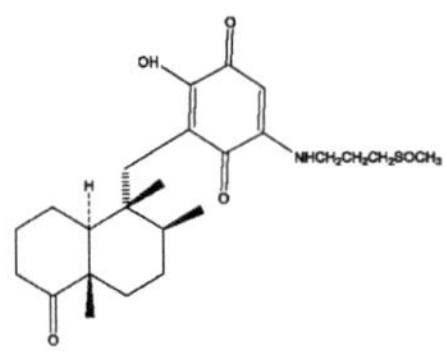	Microdilution assay *Bacillus subtilis* *Staphylococcus aureus*	MIC μM 12–5 25	[60]
Nakijiquinone V (1) Illimaquinone (2) Smenospongine (3) *Dactylospongia elegans*	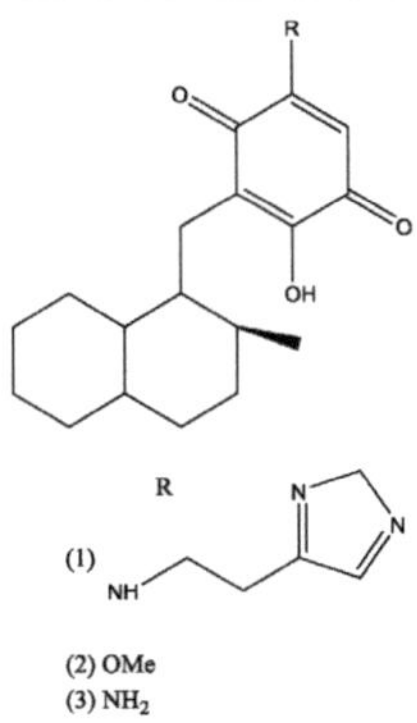	Microdilution assay *Bacillus megaterium* DSM32 *Micrococcus luteus* ATCC4698 *Escherichia coli* K12	MIC μg/mL **(1)**32 **(1)** 32 (2) 32 **(3)** NA **(1)** 64 **(2)** 32 **(3)** NA	[79]
Fugus				
Cytosporaquinone A (1) Cytosporaquinone B (2) Cytosporaquinone C (3) Cytosporaquinone D (4) *Cytospora sp*. strain CCTU A309	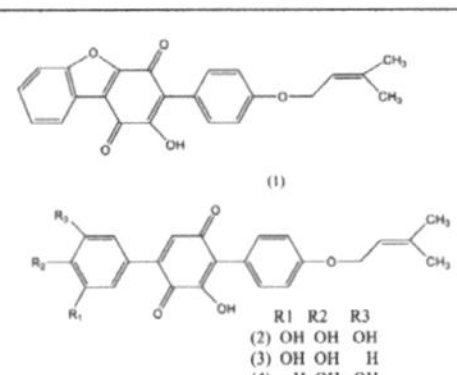	Microdilution assay *Candida albicans* DSM 1665 *Micrococcus luteus* DSM 1790 *Mucor hiemalis* DSM2656 *Rhodoturulo glutinis* DMS 10134 *Bacillus subtilis* DMS 10 *Chromobacterium violaceum* DMS 30191 *Staphylococcua aureous* DMS 346	MIC μg/mL values from 16.66 to 66.66	[80]
Auxarthrol D (1) Auxarthrol G (2) 4-hydroxyaltersolanol A (3) Altersolanol B (4) *Sporendonema casei* HDN16–802	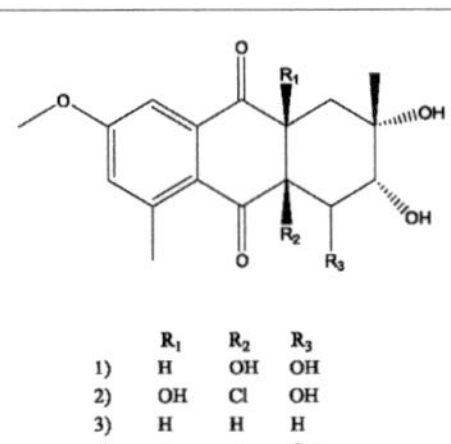	Microdilution assay *Mycobacterium phlei; Proteus sp; Bacillus subtilis; Vibrio parahemolyticus; Pseudomona aeruginosa*	MIC μM Values ranging from 12.5 to 100	[68]

Quinona/ Natural origin (plants, animals, microorganisms)	Structure	Activity or model where it was tested	Results	Reference
Bacteria				
Napyradiomycin A (1) Napyradiomycin B (2) napyradiomycin SC (3) napyradiomycin D1 (4) *Streptomyces* sp. strain CA-271078	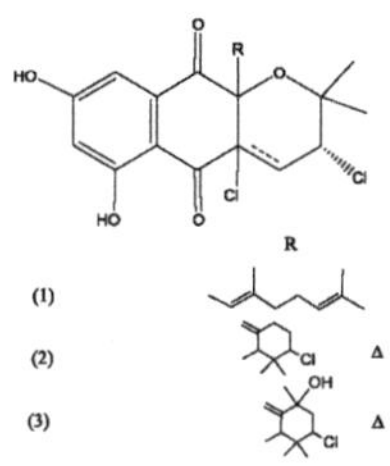 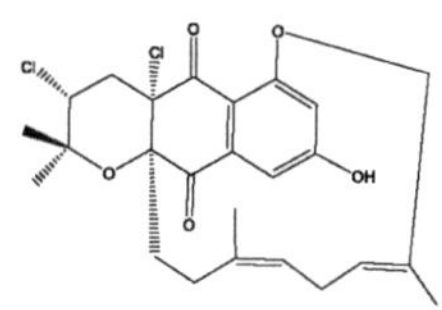 (4)	Microdilution assay Methicillin-resistant Staphylococcus aureus MB5393; *Mycobacterium tuberculosis* H37Ra	MIC μg/mL Values ranging 3–48	[72]
Animal				
3,5- dimethoxy-2-(methylthio)cyclohexa-2,5-diene-1,4-dione (1) 5-methoxy-2,3- bis (methylthio)cyclohexa-2,5-diene-1,4-dione (2) Venom of *Diplocentrus melici*	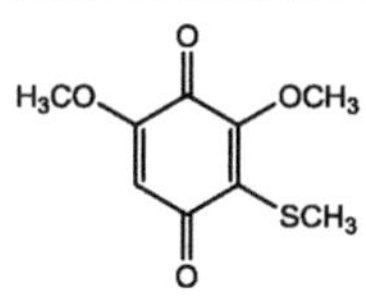 (1) 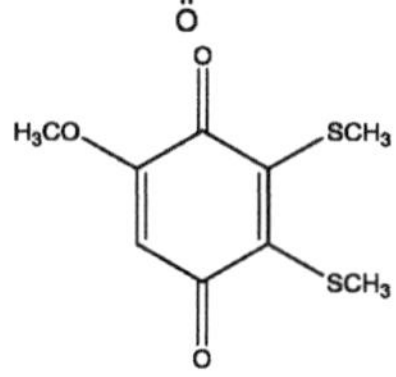 (2)	Microdilution assay *S. aureus* *M. tuberculosis*	MIC μg/mL (1) 4 (2) 6 (1) > 160 (2) 4	[81]

Minimum inhibitory concentration (MIC).

Table 2.
Quinones with antimicrobial activity.

compound to obtain the results. There are different methods to carry out these tests. In this review, the activities were determined by the use of MTT, SRB, NR, IDO, iodide propidium, violet crystal, cell counting kits, resazurin reduction, sulforhodamine B, AGS, Trypan blue, immunophenotyping, Alamar blue, FITC Annexin V Apoptosis, the CCK-8 colorimetric method, and Annexin V/7-AAD.

The determination of antimicrobial activity was carried out by MIC, micro-dilution, and broth microdilution volatilization.

Quinones have good activity against numerous cell cancer lines; they also exhibit good antimicrobial activity. This situation, along with the wide variety of structures that these compounds exhibit, make them a very interesting topic to continue to explore for other mechanisms of action and the chemical modification of their structures, among other topics.

Conflict of interest

The authors declare no conflict of interest.

Author details

Nimsi Campos-Xolalpa[1], Julia Pérez-Ramos[1], Ana Esquivel-Campos[1], Cuauhtemoc Pérez-González[1], Leonor Sánchez-Pérez[2] and Salud Pérez-Gutiérrez[1]*

1 Department of Biological Systems, Universidad Autónoma Metropolitana-Xochimilco, Calzada del Hueso 1100, Col. Villa Quietud Ciudad de México, México

2 Department of Health Attention, Universidad Autónoma Metropolitana-Xochimilco, Calzada del Hueso 1100, Col. Villa Quietud Ciudad de México, México

*Address all correspondence to: msperez@correo.xoc.uam.mx

References

[1] Khalifa SAM, Elias N, Farag MA, Chen L, Saeed A, Hegazy MF, Moustafa MS, Abd El-Washed A, Al- Mousawi SM, Musharraf SG, Chang FR, Iwasaku A, Suenaga K, Alajlani M, Göransson U, El-Seedi HR. Marine Natural products: A source of novel anticancer drugs. Mar Drugs. 2019;17:491-522. DOI: 10.3390/md17090491.

[2] Dutta S, Mahalanobish S, Saha S, Ghosh S, Sil PC. Natural products: An uncoming therapeutic approach to cancer. Food Chem Toxicol. 2019; 128:240-255. DOI: 10.1016/j.fct.2019.04.012.

[3] Gallagher EJ, Neel BA, Antoniou IM, Yakar S, LeRoith D. The increased risk of cancer in obesity and type 2 diabetes: potential mechanisms. In: Poretsky, L. Editor. Principle of Diabetes mellitus. Springer International Publishing, Cham; 2017. 731-753 pp. DOI: 10.1007/978-0-387-09841-8_36).

[4] White MC, Holman DM, Boehm JE, Peipins LA, Grossman M, Henley SJ. Age and cancer risk. A potential modificable relationship. Am J Prev Med. 2014;46:S7-S15. DOI: 10.1016/j.amepre.2013.10.029.

[5] Prestinaci F, Pezzotti P, Pantosti A. Antimicrobial resistance: A global multifaceted phenomenon. Pathog Glob Health. 2015;109:309–318. DOI: 10.1179/2047773215Y.0000000030.

[6] Morehead MS, Scarbrough C. Emergence of global antibiotic resistance. Prim Care. 2018;45:467-484. DOI: 10.1016/j.pop.2018.05.006.

[7] Morgan D, Okeke R, Laxminarayan R, Perencevich E, Weisenberg S. Non-prescription antimicrobial use worldwide: A systematic review. Lancet Infect Dis. 2011;11:692-701. DOI: 10.1016/S1473-3099(11)70054-8.

[8] Borchardt JK. The beginnings of drugs therapy: ancient Mesopotamian medicine. Drug News Perspect. 2002;15: 187-192. DOI: 10.1358/dnp.2002.15.3.840015.

[9] Saha S, Sadhukhan P, Sil P. Genistein: a phytoestrogen with multifaceted therapeutic properties. Mini Rev Med Chem. 2014;14:920-940. DOI: 10.2174/1389557514666141029233442.

[10] Asche C. Antitumour quinones. Mini Rev Med Chem. 2005;5:449-467. DOI: 10.2174/1389557053765556.

[11] Saleem M, Nazir M, Ali MS, Hussain H, Lee YS, Riaz N, Jabbar A. Antimicrobial natural products: an update on future antibiotic drug candidates. Nat Prod Rep. 2010;27: 238-254. DOI: 10.1039/b916096e.

[12] Hong-Ru W, Wei Z, Xiao-Yan P, Yuan G, Xian MU. Obulqasim, Hong-Fang L, Ying Z. Quinones and coumarins from Ajania salicifolia and their radical scavenging and cytotoxic activity. J Asian Nat Prod Res. 2015;17: 1196-1203. DOI: 10.1080/10286020.2015.1117456

[13] Li Y, Dong C, Xu MJ, Lin WH. New alkylated benzoquinones from mangrove plant Aegiceras Corniculatum with anticancer activity. J Asian Nat Prod Res. 2020;22:121-130. DOI: 10.1080/10286020.2018.1540604.

[14] Abdissa N, Gohlk, S, Frese M, Sewald N. Cytotoxic compounds from aloe megalacantha. Molecules. 2017;22: 1136-1141. DOI:10.3390/molecules22071136.

[15] Asaumi S, Kawakami S, Sugimoto S, Matsunami K, Otsuka H, Shinzato T. Alkylated benzoquinones: ardisiaquinones A–H from the leaves of Ardisia quinquegona and their anti-leishmania activity. Chem Pharm Bull.

2018;66:757-763. DOI: 10.1248/cpb.c18-00281.

[16] Yuzbasioglu Baran M, Guvenalp Z, Saracoglu I, Kazaz C, Salih B, Demirezer LO, Kuruuzum-Uz A. Cytotoxic naphthoquinones from Arnebia densiflora (Nordm.) Ledeb and determining new apoptosis inducers. Nat Prod Res. 2020;34:1669-1677. DOI: 10.1080/14786419.2018.1525714.

[17] Byeon SE, Yi YS, Lee J, Yang WS, Kim JH, Kim J, Hong S, Cho JY. Hydroquinone exhibits in vitro and in vivo anti-cancer activity in cancer cells and mice. Int J Mol Sci. 2018;19: 903-916. DOI: 10.3390/ijms19030903.

[18] Park SH, Phuc NM, Lee J, Wu Z, Kim J, Kim H, Kim ND, Lee T, Song KS, Liu KH. Identification of acetylshikonin as the novel CYP2J2 inhibitor with anti-cancer activity in HepG2 cells. Phytomedicine. 2017;15;24:134-140. DOI: 10.1016/j.phymed.2016.12.001.

[19] Chen Y, Chen ZY, Chen L, Zhang JY, Fu LY, Tao L, Zhang Y, Hu XX, Shen XC. Shikonin inhibits triple-negative breast cancer-cell metastasis by reversing the epithelial-to-mesenchymal transition via glycogen synthase kinase 3β-regulated suppression of β-catenin signaling. Biochem Pharmacol. 2019; 166:33-45. DOI: 10.1016/j.bcp.2019.05.001.

[20] Vukic MD, Vukovic NL, Djelic GT, Popovic SL, Zaric MM, Baskic DD, Krstic GB, Tesevic VV, Kacaniova MM. Antibacterial and cytotoxic activities of naphthoquinone pigments from Onosma visianii Clem. Excli J. 2017;16: 73-88. DOI: 10.17179/excli2016-762.

[21] Trivedi R, Müller GA, Rathore MS, Mishra DP, Dihazi H. Anti-leukemic activity of Shikonin: role of ERP57 in Shikonin induced apoptosis in acute myeloid leukemia. Cell Physiol Biochem. 2016;39:604-16. DOI: 10.1159/000445652.

[22] Spyrelli ED, Kyriazou AV, Virgiliou C, Nakas A, Deda O, Papageorgiou VP, Assimopoulou AN, Gika HG. Metabolic profiling study of shikonin's cytotoxic activity in the Huh7 human hepatoma cell line. Mol Biosyst. 2017;13:841-851. DOI: 10.1039/C6MB00830E.

[23] Pavan V, Ribaudo G, Zorzan M, Redaelli M, Pezzani R, Mucignat-Caretta C, Zagotto G. Antiproliferative activity of Juglone derivatives on rat glioma. Nat Prod Res. 2017;31:632-638. DOI: 10.1080/14786419.2016.1214830.

[24] Zhou YY, Guo S, Wang Y, Song HJ, Gao HR, Zhang XJ, Sun YP, Liu Y, Yang BY, Kuang HX. α-Tetralone glycosides from the green walnut husks of Juglans mandshurica Maxim. and their cytotoxic activities. Nat Prod Res. 2020; 34:1805-1813. DOI: 10.1080/14786419.2018.1561681.

[25] Zhou Y, Yang B, Jiang Y, Liu Z, Liu Y, Wang X, Kuang H. Studies on cytotoxic activity against HepG-2 cells of naphthoquinones from green walnut husks of Juglans mandshurica Maxim. Molecules. 2015;20:15572-15588. DOI: 10.3390/molecules200915572.

[26] Wu J, Zhang H, Xu Y, Zhang J, Zhu W, Zhang Y, Chen L, Hua W, Mao Y. Juglone induces apoptosis of tumor stem-like cells through ROS-p38 pathway in glioblastoma. BMC Neurology. 2017;17:70-76. DOI: 10.1186/s12883-017-0843-0.

[27] De U, Son JY, Jeon Y, Ha SY, Park YJ, Yoon S, Ha KT, Choi WS, Lee BM, Kim IS, Kwak JH, Kim HS. Plumbagin from a tropical pitcher plant (Nepenthes alata Blanco) induces apoptotic cell death via a p53-dependent pathway in MCF-7 human breast cancer cells. Food Chem Toxicol. 2019;123:492-500. DOI: 10.1016/j.fct.2018.11.040.

[28] Sameni S, Hande MP. Plumbagin triggers DNA damage response, telomere dysfunction and genome

instability of human breast cancer cells. Biomed. Pharmacother. 2016;82: 256-268. DOI: 10.1016/j.biopha.2016.05.007.

[29] Kuete V, Omosa LK, Tala VR, Midiwo JO, Mbaveng AT, Swaleh S, Karaosmanoğlu O, Sivas H. Cytotoxicity of Plumbagin, Rapanone and 12 other naturally occurring Quinones from Kenyan Flora towards human carcinoma cells. BMC Pharmacol Toxico. 2016;17:1-10. DOI: 10.1186/s40360-016-0104-7.

[30] Mancilla IA, Coatti GC, Biazi BI, Zanetti TA, Baranoski A, Marques LA, Corveloni AC, Lepri SR, Mantovani MS. Molecular pathways related to the control of proliferation and cell death in 786-O cells treated with plumbagin. Mol Biol Rep. 2019;46:6071-6078. DOI: 10.1007/s11033-019-05042-9.

[31] Ku HJ, Kwon OS, Kang BS, Lee DS, Lee HS, Park JW. IDH2 knockdown sensitizes tumor cells to Emodin cytotoxicity in vitro and in vivo. Free Radic Res. 2016;50:1089-1097. DOI: 10.1080/10715762.2016.1178739.

[32] Moreira TF, Sorbo JM, Souza FDO, Fernandes BC, Ocampos FMM, de Oliveira DMS, Arcaro CA, Assis RP, Barison A, Miguel OG, Baviera AM, Soares CP, Brunetti IL. Emodin, Physcion, and crude extract of Rhamnus sphaerosperma var. pubescens induce mixed cell death, increase in oxidative stress, DNA damage, and inhibition of AKT in cervical and Oral squamous carcinoma cell lines. Oxid Med Cell Longev. 2018;2018:1-18. DOI: 10.1155/2018/2390234.

[33] Li R, Li W, You Y, Guo X, Peng Y, Zheng J. Metabolic activation and cytotoxicity of Aloe-Emodin mediated by sulfotransferases. Chem Res Toxicol. 2019;32:1281-1288. DOI: 10.1021/acs.chemrestox.9b00081.

[34] Du Y, Zhang J, Tao Z, Wang C, Yan S, Zhang X, Huang M. Aloe emodin exerts potent anticancer effects in MIAPaCa-2 and PANC-1 human pancreatic adenocarcinoma cell lines through activation of both apoptotic and autophagic pathways, sub-G1 cell cycle arrest and disruption of mitochondrial membrane potential (ΛΨm). J BUON. 2019;24:746-753. PMID: 31128032.

[35] Zhou M, Xing HH, Yang Y, Wang YD, Zhou K, Dong W, W, Li GP, Hu WY, Liu Q, Li XM, Hu QF. Three new anthraquinones from the twigs of Cassia fistula and their bioactivities. J Asian Nat Prod Res. 2017;19:1073-1078. DOI: 10.1080/10286020.2017.1285911.

[36] Ribeiro V, Andrade PB, Valentão P, Pereira DM. Benzoquinones from Cyperus spp. trigger IRE1α-independent and PERK-dependent ER stress in human stomach cancer cells and are novel proteasome inhibitors. Phytomedicine. 2019;63:153017. DOI: 10.1016/j.phymed.2019.153017.

[37] Nordin N, Majid NA, Mohan S, Dehghan F, Karimian H, Rahman MA, Ali HM, Hashim NM. Cleistopholine isolated from Enicosanthellum pulchrum exhibits apoptogenic properties in human ovarian cancer cells. Phytomedicine. 2016;23:406-416. DOI: 10.1016/j.phymed.2016.02.016.

[38] Bigolin A, Maioral MF, Stefanes NM, Zatelli GA, Philippus AC, Falkenberg MB, Santos-Silva MC. Cytotoxic mechanisms of primin, a natural quinone isolated from Eugenia hiemalis, on hematological cancer cell lines. Anticancer Drugs. 2020;31: 709-717. DOI: 10.1097/CAD.0000000000000937.

[39] Thanuphol P, Asami Y, Shiomi K. Wongnoppavich A, Tuchinda P, Soonthornchareonnon N. Marcanine G, a new cytotoxic 1-azaanthraquinone from the stem bark of Goniothalamus marcanii Craib. Nat Prod Res. 2018;32: 1682-1689. DOI: 10.1080/14786419.2017.1396588.

[40] Zhu H, Zheng Z, Zhang J, Liu X, Liu Y, Yang W, Liu Y, Zhang T, Zhao Y, Liu Y, Su X, Gu X. Anticancer effect of 2,7-dihydroxy-3-methylanthraquinone on human gastric cancer SGC-7901 cells in vitro and in vivo. Pharm Biol. 2016; 54:285-92. DOI: 10.3109/13880209.2015.1033563.

[41] Cimmino A, Mathieu V, Evidente M, Ferderin M, Moreno Y, Banuls L, Masi M, De Carvalho A, Kiss R, Evidente A. Glanduliferins A and B, two new glucosylated steroids from Impatiens glandulifera, with in vitro growth inhibitory activity in human cancer cells. Fitoterapia. 2016;109: 138-45. DOI: 0.1016/j.fitote.2015.12.016.

[42] Feilcke R, Arnouk G, Raphane B, Richard K, Tietjen I, Andrae-Marobela K, Erdmann F, Schipper S, Becker K, Arnold N, Frolov A, Reiling N, Imming P, Fobofou SAT. Biological activity and stability analyses of knipholone anthrone, a phenyl anthraquinone derivative isolated from Kniphofia foliosa Hochst. J Pharm Biomed Anal. 2019;174:277-285. DOI: 10.1016/j.jpba.2019.05.065.

[43] Baghdadi MA, Al-Abbasi FA, El-Halawany AM, Aseeri AH, Al-Abd AM. Anticancer profiling for coumarins and related O-naphthoquinones from Mansonia gagei against solid tumor cells in vitro. Molecules. 2018;23:1020-1033. DOI: 10.3390/molecules23051020.

[44] Kuete V, Mbaveng AT, Sandjo LP, Zeino M, Efferth T. Cytotoxicity and mode of action of a naturally occurring naphthoquinone, 2-acetyl-7-methoxynaphtho [2, 3-b] furan-4, 9-quinone towards multi-factorial drug-resistant cancer cells. Phytomedicine. 2017;33:62-68. DOI: 10.1016/j.phymed.2017.07.010.

[45] Abu N, Zamberi NR, Yeap SK, Nordin N, Mohamad NE, Romli MF, Rasol NE, Subramani T, Ismail NH, Alitheen NB. Subchronic toxicity, immunoregulation and anti-breast tumor effect of Nordamnacantal, an anthraquinone extracted from the stems of Morinda citrifolia L. BMC Complem Altern M. 2018;18:18-31. DOI: 10.1186/s12906-018-2102-3.

[46] Alobaedi OH, Talib WH, Basheti IA. Antitumor effect of thymoquinone combined with resveratrol on mice transplanted with breast cancer. Asian Pac J Trop Med. 2017;10:400-408. DOI: 10.1016/j.apjtm.2017.03.026.

[47] Houdkova M, Rondevaldova J, Doskocil I, Kokoska L. Evaluation of antibacterial potential and toxicity of plant volatile compounds using new broth microdilution volatilization method and modified MTT assay. Fitoterapia. 2017;118:56-62. DOI: 10.1016/j.fitote.2017.02.008.

[48] Arumugam P, Subramanian R, Priyadharsini JV, Gopalswamy J. Thymoquinone inhibits the migration of mouse neuroblastoma (Neuro-2a) cells by down-regulating MMP-2 and MMP-9. Chin J Nat Med. 2016;14:904-912. DOI: 10.1016/S1875-5364(17)30015-8.

[49] Bowen L, Li C, Bin L, Ying T, Shijun L, Junxing D. Chemical constituents, cytotoxic and antioxidant activities of extract from the rhizomes of Osmunda japonica Thunb. Nat Prod Res. 2020;34: 847-850. DOI: 10.1080/14786419.2018.1501692.

[50] Bajpai VK, Alam MB, Quan KT, Choi HJ, An H, Ju MK, Lee SH, Huh YS, Han YK, Na M. Cytotoxic properties of the anthraquinone derivatives isolated from the roots of Rubia philippinensis. BMC Complem Altern Med. 2018;18: 200-206. DOI: 10.1186/s12906-018-2253-2.

[51] Boueroy P, Saensa-Ard S, Siripong P, Kanthawong S, Hahnvajanawong C. Rhinacanthin-C extracted from Rhinacanthus nasutus (L.) inhibits cholangiocarcinoma cell migration and

invasion by decreasing MMP-2, uPA, FAK and MAPK pathways. Asian Pac J Cancer Prev. 2018;19:3605-3613. DOI: 10.31557/APJCP.2018.19.12.3605.

[52] Boonyaketgoson S, Rukachaisirikul V, Phongpaichit S, Trisuwan K. Naphthoquinones from the leaves of Rhinacanthus nasutus having acetylcholinesterase inhibitory and cytotoxic activities. Fitoterapia. 2018; 124: 206-210 DOI: 10.1016/j.fitote.2017.11.011

[53] Dias RB, de Araújo TBS, de Freitas RD, Rodrigues ACBDC, Sousa LP, Sales CBS, Valverde LF, Soares MBP, Dos Reis MG, Coletta RD, Ramos EAG, Camara CA, Silva TMS, Filho JMB, Bezerra DP, Rocha CAG. β-Lapachone and its iodine derivatives cause cell cycle arrest at G2/M phase and reactive oxygen species-mediated apoptosis in human oral squamous cell carcinoma cells. Free Radic Biol Med. 2018;126:87-100. DOI: 10.1016/j.freeradbiomed.2018.07.022.

[54] Zada S, Hwang JS, Ahmed M, Lai TH, Pham TM, Kim DH, Kim DR. Protein kinase A activation by β Lapachone is associated with apoptotic cell death in NQO1 overexpressing breast cancer cells. Oncol Rep. 2019;42: 1621-1630. DOI: 10.3892/or.2019.7243.

[55] Zhang Q, Chen L, Hu LJ, Liu WY, Feng F, Qu W. Two new ortho benzoquinones from Uncaria rhynchophylla. Chin J Nat Med. 2016; 14:232-235. DOI: 10.1016/S1875-5364(16)30021-8.

[56] Panthong K, Hongthong S, Kuhakarn C, Piyachaturawat P, Suksen K, Panthong A, Chiranthanut N, Kongsaeree P, Prabpai S, Nuntasaen N, Reutrakul V. Pyranonaphthoquinone and anthraquinone derivatives from Ventilago harmandiana and their potent anti-inflammatory activity. Phytochemistry. 2020;169:112182. DOI: 10.1016/j.phytochem.2019.112182.

[57] Yu HB, Yin ZF, Gu BB, Zhang JP, Wang SP, Yang F, Lin HW. Cytotoxic meroterpenoids from the marine sponge Dactylospongia elegans. Nat Prod Res. 2019:1-7. DOI: 10.1080/14786419.2019.1633644.

[58] Neupane RP, Parrish SM, Neupane J, Yoshida WY, Yip ML, Turkson J, Harper MK, Head JD, Williams PG. Cytotoxic sesquiterpenoid quinones and quinols, and an 11-membered heterocycle, Kauamide, from the Hawaiian marine sponge Dactylospongia elegans. Mar Drugs. 2019;17:423-428. DOI: 10.3390/md17070423.

[59] Luo X, Li P, Wang K, de Voogd NJ, Tang X, Li G. Cytotoxic sesquiterpenoid quinones from South China Sea sponge Dysidea sp. Nat Prod Res. 2019;1-6. DOI: 10.1080/14786419.2019.1679132.

[60] Ito T, Nguyen HM, Win NN, Vo HQ, Nguyen HT, Morita H. Three new sesquiterpene aminoquinones from a Vietnamese Spongia sp. and their biological activities. J Nat Med. 2018;72: 298–303. DOI: 10.1007/s11418-017-1130-5.

[61] Yen I, Lee SY, Lin KT, Lai FY, Kuo MT, Chang WL. In vitro anticancer activity and structural characterization of ubiquinones from Antrodia cinnamomea mycelium. Molecules. 2017;22:747-752. DOI: 10.3390/molecules22050747.

[62] Li JL, Jiang X, Liu X, He C, Di Y, Lu S, Huang H, Lin B, Wang D, Fan B. Antibacterial anthraquinone dimers from marine derived fungus Aspergillus sp. Fitoterapia. 2019;133:1-4. DOI: 10.1016/j.fitote.2018.11.015.

[63] Wang M, Sun ZH, Chen YC, Liu HX, Li HH, Tan GH, Li SN, Guo XL, Zhang W. Cytotoxic cochlioquinone derivatives from the endophytic fungus Bipolaris sorokiniana derived from Pogostemon cablin. Fitoterapia. 2016;

110:77-82. DOI: 10.1016/j.fitote.2016.02.005.

[64] He W, Zhou XJ, Qin XC, Mai YX, Lin XP, Liao SR, Yang B, Zhang T, Tu ZC, Wang JF, Liu Y. Quinone/hydroquinone meroterpenoids with antitubercular and cytotoxic activities produced by the sponge-derived fungus Gliomastix sp. ZSDS1-F7. Nat Prod Res. 2017;31:604-609. DOI: 10.1080/14786419.2016.1207076.

[65] Le Pogam P, Le Lamer AC, Siva B, Legouin B, Bondon A, Graton J, Jacquemin D, Rouaud I, Ferron S, Obermayer W, Babu KS, Boustie J. Minor Pyranonaphthoquinones from the Apothecia of the Lichen Ophioparma ventosa. J Nat Prod. 2016;79:1005-1011. DOI: 10.1021/acs.jnatprod.5b01073.

[66] Zhang HM, Ju CX, Li G, Sun Y, Peng Y, Li YX, Peng XP, Lou HX. Dimeric 1,4-benzoquinone derivatives with cytotoxic activities from the marine-derived Fungus Penicillium sp. L129. Mar Drugs. 2019;17:383-389. DOI: 10.3390/md17070383.

[67] Mishra PD, Verekar SA, Deshmukh SK, Joshi KS, Fiebig HH, Kelter G. Altersolanol A: a selective cytotoxic anthraquinone from a Phomopsis sp. Lett Appl Microbiol. 2015;60:387-391. DOI: 10.1111/lam.12384.

[68] Otto C, Hahlbrock T, Eich K, Karaaslan F, Jürgens C, Germer CT, Wiegering A, Kämmerer U. Antiproliferative and antimetabolic effects behind the anticancer property of fermented wheat germ extract. BMC Complement Altern Med. 2016;1;16:160. DOI: 10.1186/s12906-016-1138-5.

[69] Ge X, Sun C, Feng Y, Wang L, Peng J, Che Q, Gu Q, Zhu T, Li D, Zhang G. Anthraquinone Derivatives from a Marine-Derived Fungus Sporendonema casei HDN16-802. Mar Drugs. 2019;17:334. DOI: 10.3390/md17060334.

[70] Li JS, Zhang H, Qi H, Wang JD, Xiang WS. Bioactive naphthoquinone and anthrone derivatives from endophytic Micromonospora sp. NEAU-gq13. J Asian Nat Prod Res. 2019;21:1151-1160. DOI: 10.1080/10286020.2018.1520222.

[71] Xu C, Sun X, Jin M, Zhang X. A novel benzoquinone compound isolated from deep-sea hydrothermal vent triggers apoptosis of tumor cells. Mar Drugs. 2017;15:200-206. DOI: 10.3390/md15070200.

[72] Carretero-Molina D, Ortiz-López FJ, Martín J, Oves-Costales D, Díaz C, de la Cruz M, Cautain B, Vicente F, Genilloud O, Reyes F. New Napyradiomycin Analogues from Streptomyces sp. Strain CA-271078. Mar Drugs. 2019;18:22. DOI: 10.3390/md18010022.

[73] Nain-Perez, A, Barbosa LC, Rodríguez-Hernández D, Kramell AE, Heller L, Csuk R. Natural abenquines and synthetic analogues: preliminary exploration of their cytotoxic activity. Bioorg Med Chem Lett. 2017;27:1141-1144. DOI: 10.1016/j.bmcl.2017.01.079.

[74] Zhou B, Jiang YJ, Ji YY, Zhang HJ, Shen L. Lactoquinomycin C and D, two new medermycin derivatives from the marine-derived Streptomyces sp. SS17A. Nat Prod Res. 2020;34:1213-1218. DOI: 10.1080/14786419.2018.1556265.

[75] Abdelfattah MS, Elmallah MI, Mohamed AA, Ishibashi M. Sharkquinone, a new ana-quinonoid tetracene derivative from marine-derived Streptomyces sp. EGY1 with TRAIL resistance-overcoming activity. J Nat Med. 2017;71:564-569. DOI: 10.1007/s11418-017-1086-5.

[76] Farooq U, Khan S, Naz S, Khan A, Khan A, Ahmed A, Riaz N. Three new anthraquinone derivatives isolated from Symplocos racemosa and their antibiofilm activity. Chin J Nat

Medicines. 2017;15:944-949. DOI: 10.1016/s1875-5364(18)30011-6.

[77] Viegas FPD, Espuri PF, Oliver JC, Silva NC, Dias ALT, Marques MJ, Soares MG. Leishmanicidal and antimicrobial activity of primin and primin-containing extracts from Miconia willdenowii. Fitoterapia. 2019;138: 104297. DOI: 10.1016/j.fitote.2019.104297.

[78] Xavier MR, Santos MMS, Queiroz MG, de Lima Silva MS, Goes AJS, De Morais Jr MA. Lawsone, a 2-hydroxy-1, 4-naphthoquinone from Lawsonia inermis (henna), produces mitochondrial dysfunctions and triggers mitophagy in Saccharomyces cerevisiae. Mol Biol Rep. 2020;47:1173-1185. DOI: 10.1007/s11033-019-05218-3.

[79] Balansa W, Mettal U, Wuisan ZG, Plubrukarn A, Ijong FG, Liu Y, Schäberle TF. A new sesquiterpenoid aminoquinone from an Indonesian marine sponge. Mar drugs. 2019;17: 158-163. DOI: 10.3390/md17030158.

[80] Narmani A, Teponno RB, Arzanlou M, Surup F, Helaly SE, Wittstein K, Praditya DF, Babai-Ahari A, Steinmann E, Stadler M. Cytotoxic, antimicrobial and antiviral secondary metabolites produced by the plant pathogenic fungus Cytospora sp. CCTU A309. Fitoterapia. 2019;134:314-322. DOI: 10.1016/j.fitote.2019.02.015.

[81] Carcamo-Noriega EN, Sathyamoorthi S, Banerjee S, Gnanamani E, Mendoza-Trujillo M, Mata-Espinosa D, Hernández-Pando R, Veytia-Bucheli JI, Possani LD,Z are RN. 1, 4-Benzoquinoneantimicrobial agents against Staphylococcus aureus and Mycobacterium tuberculosis derived from scorpion venom. Proc Natl Acad Sci U S A. 2019;116:12642-12647. DOI: 10.1073/pnas.1812334116.

Chapter 4

Some Methodological Aspects in Studies of Metal Nanoparticles' Toxicity towards Cultured Cells

Elena Mikhailovna Egorova and Said Ibragimovich Kaba

Abstract

Some actual questions arising in studies of the toxic effects of metal nanoparticle water solutions on cultured cells are considered. *First*, basic conditions required for the correct determination of nanoparticle size effect; the arguments are adduced in favor of the use of number nanoparticle concentration instead of the conventional mass one. *Second*, the problem of invalidity of the Smoluchowski equation; for charged nanoparticles the error in zeta potential value calculated from the measured electrophoretic mobility by the Smoluchowski equation cannot be neglected. *Third*, for the nanoparticles stabilized with surfactants, elucidation of the mechanism of cytotoxicity should include the determination of separate contributions of surfactant molecules and micelles into the total effect on cell viability.

Keywords: metal nanoparticles, cytotoxicity, methodological aspects, particle size effect, zeta potential, toxicity of surfactants

1. Introduction

In the last decades, the intensive development of medical applications of silver, gold and other metal nanoparticles brings into light the problem of their toxicity for a human organism. Therefore, studies on the biological activity of these nanoparticles are oriented mostly on elucidation of the mechanisms of their toxic effects on living organisms and determination of the conditions providing their safe usage. These studies lie within the scope of new branch in toxicology – nanotoxicology; general information about this direction may be found in several reviews [1–3].

One of the main lines of studies in nanotoxicology is focused on the toxicity of silver nanoparticles (AgNPs) used as water solutions. The reason lies in the widespread applications of these nanoparticles for medical purposes due mostly to their expressed antimicrobial activity; some recent results in this field are summarized in [4–6]. Studies are fulfilled mainly on the three objects: microorganisms, cultured cells and animals (mice and rats). The results obtained in the last two decades are presented in a great number of original papers, reviews and books; several examples are given in [7–9]. Nevertheless, one can conclude that there is no clearness in the main questions important for the estimation of AgNPs safety for a human organism.

Analysis of the literature as well as our long-term experience in studies of AgNPs biological effects allow to infer that the main reason lies in the field of methodology. ***First***, there is a wide variety of factors affecting the main toxicity criteria, one faces therefore the significant dispersion of their values for a given type of bacterial or mammalian cells, or for a given animal organism. ***Second***, it is not rarely met in literature that the main nanoparticle characteristics are determined incorrectly, hence the results reported are not reliable. ***Third***, the non-coordination in similar studies conducted by different laboratories takes place, so it becomes impossible, for one and the same bacterial species, cell or animal line, to obtain a sufficient pool of independent results for the nanoparticles with the same parameters, so that the reliable mean toxicity criteria values could be obtained and the corresponding safe ranges of nanoparticle concentrations established. The methodological problems under question are minutely described in our review of the cytotoxicity studies of AgNPs prepared by the biological reduction [10].

In the present chapter some methodological questions arising in studies of the AgNPs toxic effects on cultured cells are considered. Here belong (1) basic conditions required for the correct determination of nanoparticle size effect; the arguments are adduced in favor of the use of number nanoparticle concentration instead of the conventional mass one, (2) the problem with invalidity of the Smoluchowski equation used for zeta potential calculation from the measured electrophoretic mobility (EPM) of nanoparticles and (3) for the nanoparticles stabilized with surfactants, the expediency to determine separate contributions of surfactant molecules and micelles to the total effect on cell viability.

These questions are expounded in the three sections presented below. In the first two, apart from purely methodological aspects the necessary calculations are included and illustrations on examples taken from literature which make clear the significance of the corrections discussed. In the last section we briefly describe our original approach suggested for gaining a more detailed information about the mechanism of cytotoxicity in case when nanoparticles are stabilized with surfactant. The approach is illustrated on example of anionic surfactant–stabilized AgNPs interaction with Jurkat cells.

2. Basic requirements for the studies of particle size effect on living cells

Studies on the biological activity of metal nanoparticles are devoted mostly to estimation of the influence of one or another nanoparticle characteristic on functional activity of bacterial or mammalian cells and multicellular organisms. Up till now the majority of works were carried out with AgNPs. This was conditioned, *first*, by the relatively simple ways of preparation and nanoparticle stability and *second*, by the wide possibilities of their application as antimicrobial means, both as water solutions and as small additives to various liquid-phase or solid materials.

The main characteristics of nanoparticle parameters are size, form, surface charge and the nature of stabilizer; the latter two are often referred to as surface properties. In studies of the biological activity the nanoparticles are used as their water solutions; the solutions should satisfy to the general conditions considered in our monograph [11]; here we described also the criteria used for the quantitative estimation of the influence of nanoparticle parameters on viability of the biological objects, as well as results of the relevant investigations on microorganisms, including some methodological problems.

In the more recent textbook [12] we considered the basic methodological requirements for experimental design in studies on the influence of the main nanoparticle parameters mentioned above. This point deserves attention because the nanoparticles as biologically active agents are principally different from molecular solution of an individual substance, namely, a nanoparticle solution is a complex object, composed of several components, including those capable of exerting their own effect. This distinction should be particularly emphasized, since, for the time being, its importance is not always realized by the researchers (mostly biologists) using ready-made nanoparticle solutions either supplied by the nanotechnological companies or synthesized by the other laboratories. Estimation of the nanoparticle effects is carried out often according to the standard protocols accepted for the effects of individual substances in water solutions. It is clear however that, to provide the correct experimental procedure in studies of the nanoparticle cytotoxicity, it is necessary to take into account the properties of the nanoparticles as special system so as to obtain the results which really allow to achieve the goal pursued in experiment.

This is true even in case of the simplest task – a study of the nanoparticle concentration effect on various properties of cells. The more so it is justified for the special requirements in studies of the influence of basic nanoparticle parameters on the characteristics of living cells' state. As seems evident from methodological point of view, to determine the influence of one main nanoparticle parameter, the three others should remain constant. Here the importance of this rule is illustrated on example of the nanoparticle size effect on cell viability.

To estimate in experiment the toxic effect of particle size, one should eliminate the toxic action of the three other nanoparticle parameters, that is, to create the conditions where a cell response registered is not sensitive to the changes of particle form, surface charge or surface composition. These conditions are formulated below:

1. **Nanoparticles of different (no less than two) sizes in the range 1–100 nm should have the same form, surface charge and (for those obtained with stabilizer) the same composition of a stabilizing shell;**

2. **In case if the nanoparticles of different sizes are prepared by means of synthesis in the water medium, it is desirable that they were synthesized by the same procedure, since this provides the minimal difference in the composition of nanoparticle solutions (and hence, in the possible side-effects);**

3. **Size distribution of nanoparticles with different mean size should be narrow enough to exclude overlapping;**

4. **The action of nanoparticles with different sizes should be studied by the same method, on the same cells and at the same cell concentrations**.

If these conditions are satisfied, cell viability dependence on the concentration of nanoparticles with different sizes allows to estimate the difference in viability criteria and thus to clear out, which size of the nanoparticles under question is more toxic for the cells studied. It is important to stress here that, in studies of the size effects the necessity arises to change the conventional way of expressing the nanoparticle concentration and hence the experimental design. This statement was briefly substantiated earlier in our monograph ([11], p. 187). In view of its significance for the correct determination of the connection between nanoparticle size and cytotoxicity and for the corresponding applications of nanomaterials we thought it reasonable to dwell upon this question in more detail.

In all the relevant publications available, the biological activity of differently sized nanoparticles is compared by determination of the dose-effect dependence, where under "dose" one means the mass nanoparticle concentration, expressed in mg/L or μg/mL. Meanwhile, for the correct solution of the question about the influence of particle size a cell viability dependence should be measured not on the mass, but on the number and (or) "surface" nanoparticle concentration, that is, on the number of nanoparticles or on their total surface per unit volume of solution under study. This becomes clear if attention is payed to the fact that the influence of nanoparticle size proper manifests itself after nanoparticle adsorption on a cell surface. Then, by analogy with the well-known regularities of molecular adsorption on solid surfaces, it is natural to believe that the adsorption density (the number of adsorbed nanoparticles per unit of cell membrane surface) will be proportional to their number nanoparticle concentration in solution (in a cell medium used in experiment). In other words, **the effect of nanoparticles' size depends on their number concentration in cell surrounding medium**. Since at the same mass concentration, the nanoparticles of different sizes have different number concentrations, estimation of the size effects should be carried out by comparison of a cell response (e.g. viability) for the same *number* nanoparticle concentrations.

Besides, when the nanoparticles are introduced into solution (cell medium), metal ions are released from the nanoparticle surface, the ions concentration in solution volume being proportional to the "surface" nanoparticle concentration defined above. Thus, investigation of the biological activity of nanoparticles having different sizes allows to find the contributions of the two different mechanisms of their action – *first*, the effect of size proper and, *second*, the effect of metal ions released from the nanoparticle surface. However, when a cell response on the *mass* nanoparticle concentration is measured, both mechanisms act simultaneously, so there is no possibility to separate the size effect from that of metal ions. To separate the two mechanisms, one should measure the dependences of cell response either on number nanoparticle concentration (the effect of nanoparticle size) or on their "surface" concentration (the effect of metal ions).

As a result, the necessity arises to supplement the four conditions mentioned above with the fifth one, which should be fulfilled in order to obtain the exact answer to the question: "How the nanoparticle size affects its biological activity?" This additional condition can be formulated as follows:

5. **To estimate the influence of nanoparticle size one should obtain the dependence of cell reaction on the number, but not mass concentration of differently sized nanoparticles in cell incubation medium.**

As far as we know, for the time being, this condition was not met in studies of the nanoparticle size effects, simply because nobody was aware of the importance of these considerations. Unfortunately, almost the same is true for the necessity to keep to the equality of the three other nanoparticle parameters (***condition 1***), since one of them – particle surface charge – in principle cannot be estimated correctly from zeta potentials obtained using the devices based on photon correlation spectroscopy (PCS) technique. The problem here is that, to our knowledge, in most cases the software installed in these instruments uses the Smoluchowski equation for zeta potential calculation from the measured EPM. However, as shown by the more general theories of electrophoresis developed in the past century [13–20], there are practically significant combinations of conditions (small particle size, not very small charge, and low ionic strength of the medium) where the assumptions used in the Smoluchowski theory do not work. As shown in the review [21] and explained also in our monograph ([11], p. 191) the Smoluchowski theory is not valid for the charged nanoparticles in the whole nanodimensional range (1–100 nm). Therefore, zeta potential values found by the Smoluchowski equation cannot be regarded as correct measure of the nanoparticle charge and the more general theories should be applied. A detailed discussion of this point is given in the next section.

Apart from this, in some cases one faces the impossibility to obtain the nanoparticle samples with narrow enough size distribution at the small difference of mean particle sizes, hence the overlap of standard deviations takes place, i.e., the ***condition 3*** is also violated. Consequently, even if the other conditions are satisfied, the overlap of standard deviations leads to the non-reliable conclusions. As an example, we consider the work by Panacek et al. [22]. The authors studied the effect of AgNPs of different sizes on several bacterial species. As follows from experimental section, the nanoparticles were synthesized by the same technique; 4 nanoparticle samples were used, of the same (approximately spherical) form with mean sizes of 25, 35, 44 and 50 nm and narrow distribution. The antibacterial activity was determined by the same method on the same cells. So, the conditions 1–4 were satisfied and even the equality of surface charge seems to be fulfilled. The authors concluded that "the 25 nm-sized silver particles synthesized via reduction by maltose showed the highest activity, comparable even with ionic silver for certain strains" ([22], p.16252). Thus, one can infer that, the smaller are the nanoparticles, the higher is their biological activity. However, examination of the mean particle sizes and standard deviations shows that all the nanoparticle samples have too wide distributions and the standard deviations overlap, so that it is hardly possible to regard the samples studied as different in sizes and hence, to claim that the difference in particle sizes is responsible for the difference in their antibacterial activity.

In the publication discussed above, as well as in some other *in vitro* studies of AgNPs interaction with cells (e.g. [23–30]) the authors came to the conclusion that, among various sizes explored, the most toxic are small nanoparticles (less than 25 nm in size). However, there are grounds to doubt the reliability of such conclusions, because in these studies both the condition of equality of the three other nanoparticle parameters (in what concerns the surface charges), and that of the equality of number nanoparticles concentration (***condition** 5*) are sure to be violated. The importance of the last condition may be elucidated using the results of calculations of number and surface concentrations for AgNPs in a chosen size range at the same mass concentration.

For a monodisperse suspension of AgNPs at their mass concentration 10^{-3} mol/L (108 μg/mL) the number concentration of particles in 1 L of solution (N_{tot}) may be found from the general formula (see for details Appendix in [31]):

$$N_{tot} = N_A \cdot 10^{-3} / N_{Ag} = 6.023 \cdot 10^{23} \cdot 10^{-3} / 41.84 \cdot (d_{NP})^3 \quad (1)$$

Where N_A is the Avogadro number, N_{Ag} is the number of atoms in a single silver nanoparticle, d_{NP} is particle diameter (nm). Since the nanoparticles are introduced to a cell culture usually to the mass concentrations in the range 1–100 μg/mL, the corresponding number concentrations are expressed in the number of particles per 1 mL. For example, for the size of 5 nm and mass concentration, C_M = 1 μg/mL, the number concentration, C_N will be found as:

$$C_N = \left[6.023 \cdot 10^{23} \cdot 10^{-3} / 41.84 \cdot (5)^3\right] \cdot 10^{-3} / 108 = 1.066 \cdot 10^{12} \text{ particles / mL.} \quad (2)$$

The surface concentration or the total surface area (C_S) for the nanoparticles of diameter d_{NP} is found as the surface area of a single particle (S_{NP}) multiplied by the number of particles in 1 mL of suspension:

$$C_S = S_{NP} \cdot C_N (d_{NP}) = \pi \cdot (d_{NP})^2 \cdot C_N \left(m^2/mL\right) \quad (3)$$

As follows from **Table 1**, at the equal mass nanoparticle concentration, the increase of particle diameter from 5 to 100 nm leads to the decrease of their number concentration, C_N, by 4 orders. This may be illustrated by the C_N (d_{NP}) dependence

Particle size, nm	Number concentration, C_N (particles/mL)	Surface concentration, C_S, m^2/mL
5	$1.064 \cdot 10^{12}$	$8.4 \cdot 10^{-5}$
10	$1.33 \cdot 10^{11}$	$4.18 \cdot 10^{-5}$
15	$3.94 \cdot 10^{10}$	$2.785 \cdot 10^{-5}$
20	$1.066 \cdot 10^{10}$	$2.09 \cdot 10^{-5}$
25	$8.512 \cdot 10^{9}$	$1.67 \cdot 10^{-5}$
30	$4.9 \cdot 10^{9}$	$1.38 \cdot 10^{-5}$
50	$1.064 \cdot 10^{9}$	$8.36 \cdot 10^{-6}$
70	$3.9 \cdot 10^{8}$	$6.0 \cdot 10^{-6}$
100	$1.33 \cdot 10^{8}$	$4.18 \cdot 10^{-6}$

Table 1.
Number and surface concentrations of 5–100 nm AgNPs in solutions with nanoparticle concentration 1 μg/mL.

in the range 5–30 nm (**Figure 1**). It is seen that the sharpest fall of number concentration takes place at small sizes (below 10 nm). Hence it is clear that comparison of the biological activity of nanoparticles with sizes of, say, 5 and 20 nm, at such a significant difference in their number concentrations does not allow to obtain the reliable data on the influence of the size proper. Similarly looks the corresponding dependence of surface concentrations, $C_S(d_{NP})$ (**Figure 2**). Here the decrease of specific particle surface in the same size range is more smooth than in the case of number concentrations and remains within one order of magnitude; still it should be taken into account if it is important to know the contribution of Ag^+ ions released from the nanoparticle surface into the nanoparticles action upon cells.

The effect of nanoparticle size at the equal number concentration can be demonstrated using some literary experimental data obtained in studies of the action of differently sized nanoparticles on viability or other cell parameters. In case of viability, the measured viability values, V_c, after incubation with nanoparticles should be

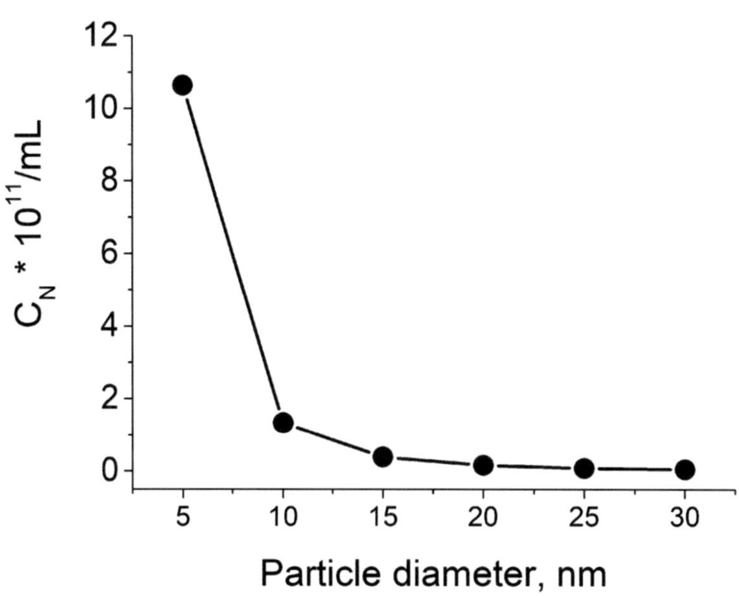

Figure 1.
*Dependence of AgNPs number concentration (C_N) on the nanoparticle size in the range 5–30 nm (data from **Table 1**).*

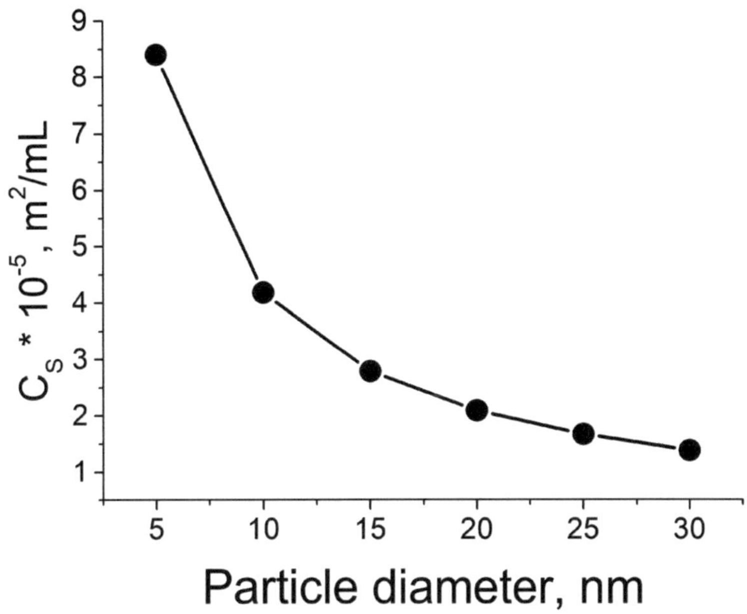

Figure 2.
Dependence of AgNPs surface concentration (C_S) on the nanoparticle size in the range 5–30 nm (data from ***Table 1****).*

compared at the equal number nanoparticle concentration. Such comparisons were found possible, for example, for the data reported in [25], where the V_c dependences on mass concentration for AgNPs of different sizes were measured in the wide range of mass concentrations so that one could find the suitable pairs of V_c values.

The authors used the citrate-stabilized AgNPs, 10, 20, 50 and 100 nm in size, produced by nanoComposix (San Diego, USA). The particle sizes were found by PCS and TEM, electron micrographs of good quality **(Figure 3)**, mean diameters were close to those claimed by the manufacturer, standard deviations narrow enough, so the nanoparticle sizes are quite reliable. It is clear also that all the nanoparticles are

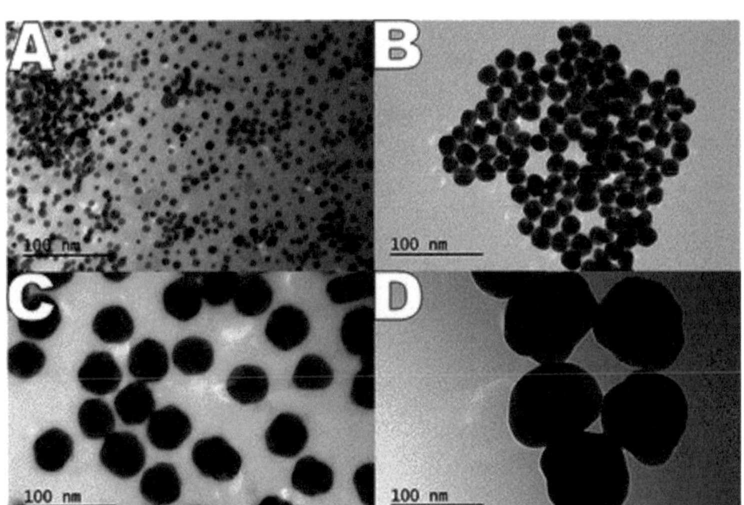

Figure 3.
TEM micrographs of AgNPs used in all experiments discussed: A – 10 nm; B – 20 nm; C – 50 nm; D – 100 nm. Reproduced from [25].

spherical in form and negatively charged, as follows from their negative zeta potentials. Apart from the other experimental data the authors present the cell viability dependences on mass nanoparticle concentration; the latter was no higher than 50 μg/mL (**Figure 4**). The Jurkat and THP-cells taken in equal concentrations ($2 \cdot 10^5$/mL) were incubated with nanoparticles for 24 hours. In both cases the viability was estimated from the mitochondrial activity by means of MTS test. Thus, the experimental design satisfied the ***conditions 1–4***, except for the equality of surface charge values; this condition was probably violated due to the unreliable zeta potential values, the defect common to practically all studies performed with nanoparticles in the last decades.

To compare the toxic effect for the nanoparticles of different sizes we calculated the number concentrations of 10 and 20 nm AgNPs at various mass concentrations (**Table 2**) and so obtained the mass 20 nm nanoparticles' concentrations equal or nearly equal to those of 10 nm nanoparticles. As seen from the table, for 20 nm AgNPs there are 4 C_N values corresponding to those of 10 nm nanoparticles at their concentrations in the range 1–6 μg/mL. For these C_N values from **Figure 4** the V_c values have been found at the four mass concentrations (8, 30, 40 and 50 μg/mL)

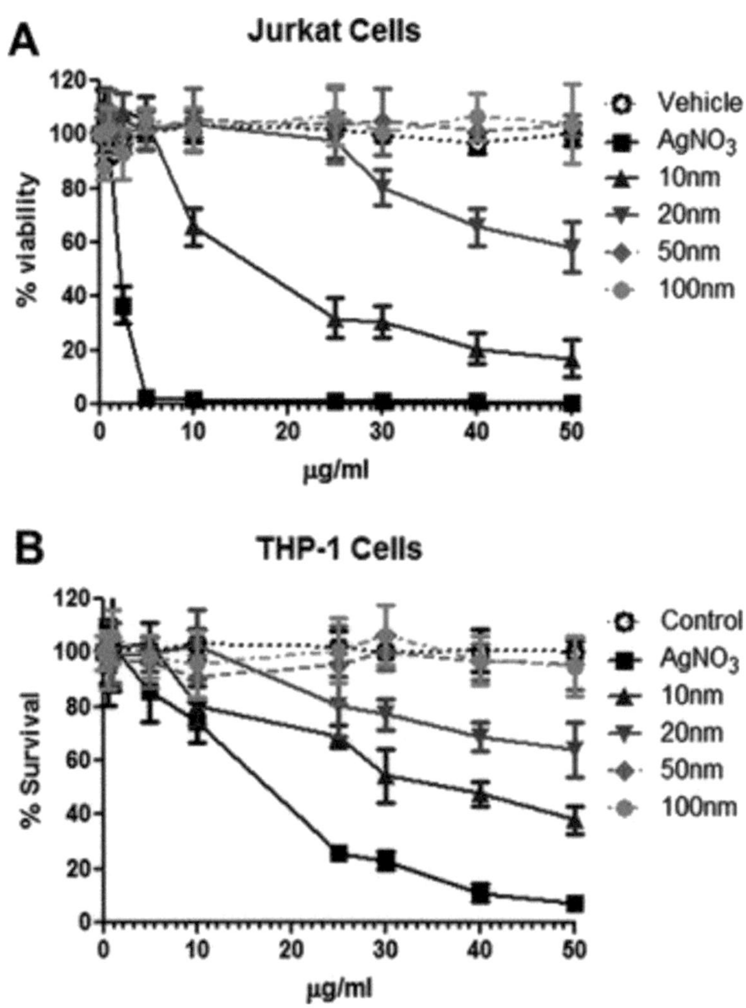

Figure 4.
The mitochondria activity of Jurkat (A) and THP-1 (B) cells after 24 h incubation with various concentrations of 10 nm, 20 nm, 50 nm, 100 nm AgNPs and with the corresponding $AgNO_3$ concentrations. Reproduced from [25].

Mass concentration C_M, μg/mL	C_N, particles/mL d = 10 nm	C_N, particles/mL d = 20 nm
1	**$1.33 \cdot 10^{11}$**	$1.66 \cdot 10^{10}$
2	$2.66 \cdot 10^{11}$	$3.32 \cdot 10^{10}$
3	$4 \cdot 10^{11}$	$5 \cdot 10^{10}$
4	**$5.32 \cdot 10^{11}$**	$6.64 \cdot 10^{10}$
5	**$6.65 \cdot 10^{11}$**	$8.3 \cdot 10^{10}$
6	**$8 \cdot 10^{11}$**	$9.96 \cdot 10^{10}$
7	$9.31 \cdot 10^{11}$	$1.16 \cdot 10^{11}$
8	$1.06 \cdot 10^{12}$	**$1.33 \cdot 10^{11}$**
10	$1.33 \cdot 10^{12}$	$1.66 \cdot 10^{11}$
20	$2.66 \cdot 10^{12}$	$3.32 \cdot 10^{11}$
30	$4 \cdot 10^{12}$	**$5 \cdot 10^{11}$**
40	$5.32 \cdot 10^{12}$	**$6.64 \cdot 10^{11}$**
50	$6.65 \cdot 10^{12}$	**$8.3 \cdot 10^{11}$**

The equal or similar C_N values at different nanoparticle sizes are marked in bold.

Table 2.
Number concentrations (C_N) of 10 and 20 nm AgNPs at various mass concentrations.

for the 20 nm nanoparticles. The results are shown in **Figure 5**. It turned out that, for the two cell types studied, the V_c values for 20 nm nanoparticles were less than 100% and decreased with the increase of C_N. At the same time, the corresponding V_c values for 10 nm particles remained at the 100% level, i.e., demonstrated the absence of toxicity. Hence follows that, at the equal number concentration the bigger nanoparticles possessed a higher toxicity than the smaller ones.

Similar conclusion can be drawn from the results reported in several other studies on the size dependence of AgNPs toxicity conducted on various cell types [26–30]. **Table 3** presents the experimental data that could be extracted from the plots of cell response vs. mass nanoparticle concentration. Similarly to what has been done with the data reported by Butler et al. [25], we have chosen the pairs of experimental values obtained for the same number concentration of differently sized nanoparticles at the relevant mass concentrations within the range studied. As seen from the table, for different combinations of sizes and cell lines, the bigger nanoparticles exert a stronger toxic influence at least on the two kinds of cell response: V_c and RCD (relative cell population doubling). It should be noted that, at present, there is a very limited number of data that can be used for the comparison under question. The main restriction here is that, in the majority of publications available, mass concentration of the bigger nanoparticles required for their number concentration to be equal to that of the smaller ones exceeds the upper limit of mass concentration in the range studied.

Basing on our calculations and analysis of the published results presented above one may suppose that, at the equal number nanoparticle concentration, the increase of particle size leads to the enhancement of its biological activity. If this supposition is confirmed in future experiments (at least for AgNPs), this will lead to the conclusion opposite to that present in the majority of the known studies of the nanoparticle size effects on animal or bacterial cells. Certainly, for the final statement the transfer from mass to number concentrations is not sufficient. The true zeta potential values should also be known, found from the measured electrophoretic

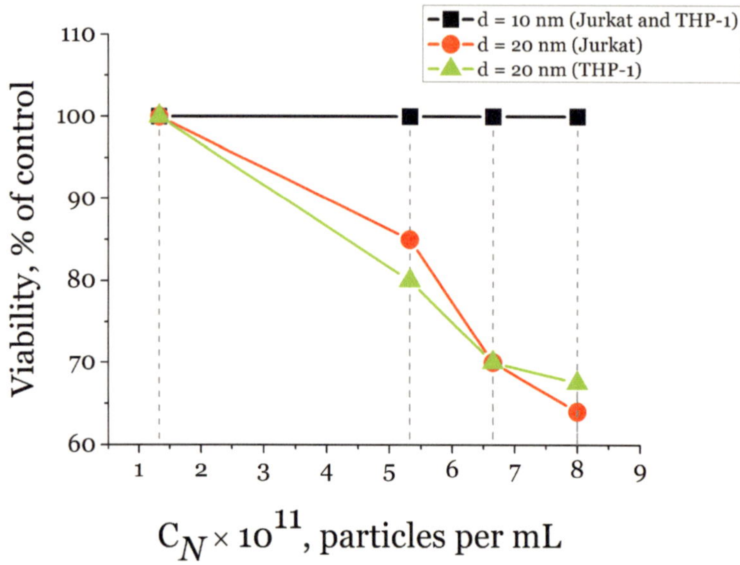

Figure 5.
Viabilities of Jurkat and THP-1 cells at the four relevant number concentrations of 10 nm and 20 nm AgNPs taken from the **Table 2**. *The viability values are elicited from the data shown in* **Figure 4**. *See text for details.*

D_{NP} (1), nm	D_{NP} (2), nm	C_N, particles/mL	C_M (1–2) µg/mL	Cell type/ Stab	Tox (1), %	Tox (2), %	Ref.
10	30	$1.33 \cdot 10^{11}$	1–27	Jurkat/ /PEI /PEG /citrate	V_c 100	V_c 10 30 45	[26]
20	50	$\approx 3.3 \cdot 10^{10}$	2–30	MLC	RPD 85	RPD 30	[27]
20	40	$4.17 \cdot 10^{11}$	25–200	THP-1/ tripeptide (Lys-Lys-Cys)	83	50	[28]
20	50	$1.67 \cdot 10^{11}$	10–150	RAW 264.7/ bovine serum albumin	60	10	[29]
5	50	$1.07 \cdot 10^{11}$	0.1–100	HaCaT/ PVP	100	60	[30]

Abbreviations: C_N – number concentration, C_M – mass concentration, V_c – cell viability.
Stab – stabilizer, PEI – polyethylene imine, PEG – polyethylene glycol, MLC – mouse lymphoma cells, PVP – polyvinyl pyrrolidone, RAW 264.7 - mouse monocyte macrophage cell line, Tox – parameter characterizing the AgNPs toxicity; RPD – relative population doubling.

Table 3.
AgNPs toxicity for cells obtained from experimental data reported in literature for the two different nanoparticle diameters (D_{NP}) at the equal number concentrations. Incubation for 24 h.

mobilities (EPM) by the equations considering the relaxation effect. The problem with invalidity of the Smoluchowski equation in studies of metal nanoparticles is discussed in the next section.

3. The problem with invalidity of the Smoluchowski equation in determination of particle surface charge using zeta potential calculated from the measured electrophoretic mobilities

As mentioned above, particle surface charge is one of the basic parameters used in studies of the biological action of metal nanoparticles. The sign and magnitude of surface charge may be found from electrokinetic (or zeta, ζ) potential calculated from the measured EPM of nanoparticles. For this purpose, the well-known relations of the classical electrical double-layer (EDL) theory may be used, modified conformably to the studies of membranes charged due to the presence of ionized groups (e.g. liposomes or biological vesicles) [32, 33]. Here belong also metal nanoparticles coated with surfactants or polymers bearing ionizable groups.

In studies of the biological activity of nanoparticles, their surface charge is characterized almost exclusively by ζ-potential values and not by those of surface charge density, and this is considered to be sufficient for the estimation of nanoparticle charge effects on functional activity of living cells or other objects explored. Therefore, we dwell here upon the methodological questions essential for determination just of ζ-potential; for the readers interested in estimation of surface charge the above references may be recommended, as well as some works where the nanoparticle surface charges are reported (e.g. [34]).

In this section we give a brief account of theoretical considerations necessary, to our view, for the correction of possible errors in determination of ζ-potential magnitude. The necessity of such an account issues from the fact that, as far as we know, at present the majority of researchers dealing with metal or metal oxide nanoparticles operate with ζ-potentials obtained by PCS technique on the devices which do not allow to obtain the correct ζ-potential values. As we noted earlier ([11], p. 190)," the problem is that standard software incorporated in the corresponding instruments (manufactured by Malvern, Coulter Electronics, or Brookhaven) usually calculates ζ-potential from the EPM measured by PCS without the relaxation correction, applicable for the measurements of nanoparticles. At the same time, the correction is essential in view of the small particle size, because a substantial error in the ζ-potential value is otherwise inevitable, as was shown, for instance, in [35–37]. Such errors are likely to occur in the relevant studies since the ζ-potential is usually calculated from the Smoluchowski equation, and the resulting values (ζ_{sm}) may considerably differ from the true ζ-potential".

"The relaxation correction" under question appears as a consequence of invalidity of the Smoluchowski equation in experimental conditions which do not satisfy the two basic assumptions of the Smoluchowski theory: (1) the width of EDL is small compared to the particle radius and (2) the double layer near the charged surface is in equilibrium state. The first assumption is determined by the inequation:

$$\kappa\alpha >> 1 \tag{4}$$

where κ is reciprocal Debye length, α is particle radius. The second assumption allows interpretation of the ζ_{sm} values within the frames of classical double-layer theory. These assumptions issue from the notion that the force applied to a particle by the external field is equilibrated by (a) viscous resistance of the fluid and (b) the electroosmotic motion of the fluid along the particle surface. Thus, here the electrophoretic retardation is considered, similar to that taking place at the movement of ion surrounded by the ionic atmosphere, but the influence of ionic atmosphere deformation at the particle movement (known as the relaxation effect) is neglected (e.g. [16]). The neglection of the ionic atmosphere deformation follows from the assumed equilibrium of the EDL. However, this assumption can be invalid even for the thin double layer determined by the expression (4). This can manifest itself in experiment as EPM dependence on particle size, the possibility which is not considered by the Smoluchowski theory (e.g. [16, 20, 38]).

According to the views developed in the electrophoresis theory, the relaxation effect is conditioned by the two main causes: (1) by the existence of surface conductivity, since the ionic surface currents change the potential distribution in the vicinity of a particle and (2) by double-layer deformation (polarization) under the influence of external field [16, 20]. Both phenomena are interconnected and represent a complicated combination of processes proceeding in the double layer of a moving particle. In the past century several theoretical approaches were suggested for the description of double layer relaxation; the most widely known are those developed by Overbeek-Booth-Wiersema [14, 15], O'Brien and White [18, 19] and S.S.Dukhin and co-workers [16, 17].

To illustrate the necessity of correction for the relaxation effect, we give here one example from the results of our PCS measurements of ζ- potential dependences on ionic strength of salt solution obtained for liposomes made from negatively charged phospholipids (**Figure 6**). The measurements were conducted in monovalent electrolytes in the range 10^{-3} – 1 M. It was established that, at the low ionic strengths, ζ-potentials calculated from the measured EPMs by the Smoluchowski equation (ζ_{sm}) differ significantly from the true zeta potential (ζ) calculated from the classical EDL theory equations [35–39]. It is clear that the deviation can be essential even for the particles with diameter of 400–600 nm (0.4–0.6 μm), especially at low salt concentrations; for example, at C = 10^{-3} M the difference $\zeta - \zeta_{sm}$ exceeds 100 mV. It is seen also that much more satisfactory agreement with experimental ζ_{sm} values is achieved with the use of the equation suggested by the Dukhin theory [34, 35, 40] which allows for the relaxation effect.

The results of such studies brought us to the conclusion that, for the correct determination of zeta potential for liposomes in the wide range of the meaningful parameters – particle size, charge, and ionic strength of water solution – for the calculation

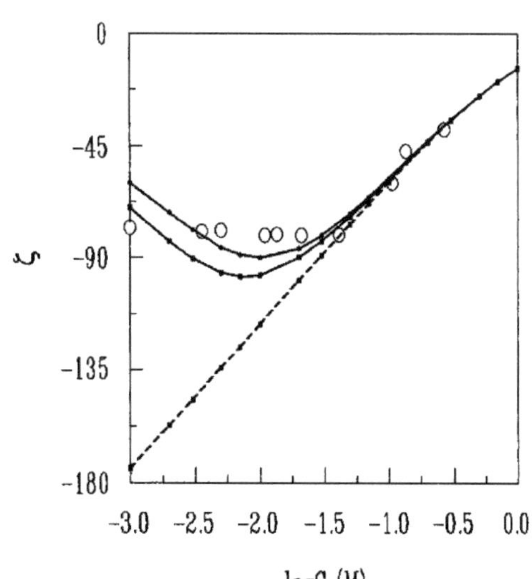

Figure 6.
ζ_{sm} as a function of electrolyte concentration for phosphatidylserine liposomes in NaCl. Open circles – experimental data. Solid lines – theoretical predictions for ζ_{sm} values calculated from one of the equations suggested by the Dukhin theory [35, 36] for the known membrane charge and liposome size 0.4 μm (upper curve) and 0.6 μm (lower curve). Dashed line – the corresponding change of true ζ-potential predicted by the classical EDL theory. Reproduced from [39, 41].

of true ζ-potential from the measured EPM one should apply not the Smoluchowski formula, but the more general relations between EPM and ζ-potential suggested by the Dukhin theory. It became clear also that correction for the relaxation effect is important for the estimation of ζ-potential (and hence, of particle surface charge) not only for lipid membranes, but for the wide range of objects studied both in colloid chemistry (e.g. latexes) and in membrane biophysics (biological vesicles, small cells et al) and in any other fields where small charged particles are explored. At the same time, since the due regard for the relaxation effect is necessary not in each case, but only for the definite combinations of the three meaningful parameters, it was of practical importance for the researchers to clear out, first of all, whether the Smoluchowski formula was applicable in the given experimental conditions. That is, whether it is reasonable to use for ζ-potential calculation any of the more complicated relations absent in the instruments applied for the EPM measurements under consideration.

To facilitate the solution of this question in studies of lipid membranes, we suggested a simple approach based on the calculation of criterion Rel[1] used in the S.S.Dukhin theory for the estimation of significance of the relaxation effect [37]:

$$\text{Rel} = \left[\exp\left(\bar{e}|\psi_d|/2kT\right) - 1\right]/\kappa\alpha \quad (5)$$

where ψ_d is Stern potential here assumed as equal to the surface potential ψ_s (see e.g. [32]), κ is reciprocal Debye length, α is particle radius. For the high charges and low ionic strengths this equation is reduced to

$$\text{Rel} = \left[\exp\left(\bar{e}|\psi_d|/2kT\right)\right]/\kappa\alpha \quad (6)$$

If electrophoresis is not complicated with surface conductivity and is not sensitive to the form of particles, that is, the Smoluchowski formula is valid, then Rel should satisfy the condition:

$$\text{Rel} << 1 \quad (7)$$

As shown in [37], for the sizes, surface charges and ionic strengths taking place in lipid membrane studies, condition (7) is satisfied when $0.04 < \text{Rel} < 0.06$. As for the nanoparticles, it is easy to guess that, for the corresponding particle sizes this condition cannot be fulfilled, and therefore the Smoluchowski equation is, in principle, invalid. The same conclusion was made recently in the review of the methods used for zeta potential determination [21]. As an illustration we give here an example of the calculation of criterion Rel for the nanoparticles stabilized with anionic or cationic surfactant bearing dissociating groups fully ionized at the neutral pH in the diluted salt solution; the relevant expressions for the calculation of κ and φ_s may be found elsewhere [37, 39, 41]. In this case surface charge density of the particles is maximal: $\sigma = \sigma^{max}$. For the area per surfactant molecule in the dense monolayer on the particle surface equal to 1ē/60 $Å^2$, $\sigma^{max} = 26.67$ μCoul/cm^2, $\varphi_s = \varphi_d = -252.7$ mV. For the mean particle diameter 20 nm (i.e. radius = 10 nm = 100 Å) and ionic strength of solution, $C = 10^{-3}$ M, we obtain: 1/κ = 96.34 Å, κ = 0.0104 1/Å, κα = 1.04. Then Rel = [exp (252.7/50.86)] /1.04 = **138.2 > > 1.**

For the nanoparticles with diameter of 100 nm Rel = **27.64 > > 1**. For $C = 10^{-2}$ M and particle size 100 nm we obtain $\varphi_s = -194.2$, 1/κ ≈ 30 Å, κα = 16.7, Rel = [exp

[1]This criterion has also the other name (Du, abbreviation from Dukhin), suggested by Lyklema [41]. Here we use the initial name Rel, accepted in the fundamental works of the author, S.S.Dukhin and in the subsequent publications of his followers.

(194.2/50.86)] /16.7 = **2.73** > **1**. Thus, even for the particle size on the upper boundary of nanometer range at ionic strengths of the order of millimoles the calculation of zeta potential by the Smoluchowski equation leads inevitably to the erroneous results. It is clear that, the higher is particle charge and the smaller its diameter and the lower is ionic strength, the bigger is the error.

In summary one can conclude that zeta potential values calculated from the measured EPMs according to the Smoluchowski equation used in the PCS devices mentioned above are not equal to the true zeta potential (ζ) determined within the frames of classical EDL theory and therefore cannot be applied for the estimation of particle surface charge density. Actually, a researcher obtains here only the equivalent of EPM calculated by the Smoluchowski equation and expressed in potential units (ζ_{sm},); this value may be essentially smaller than the true zeta potential, that is, $\zeta_{sm} < \zeta$ in a wide range of particle sizes and charges as well as of salt concentrations in the nanoparticle solution. Therefore, the correct zeta potential values for nanoparticles can only be obtained if one applies either the more general analytical expressions developed in the theory of electrophoresis [13, 14, 16, 17, 41] or the numerical methods [15, 18–20], According to our experience in the field, for highly charged particles an optimal way is suggested by one of the equations of the Dukhin theory of electrophoresis [36, 40, 41].

4. On the mechanism of cytotoxicity for the nanoparticles stabilized with surfactant

As may be deduced from the literature on the biological activity of metal or metal oxide nanoparticles, in the last decade the data has been reported, testifying to the significance of one more meaningful parameter, namely, of the activity manifested by the nanoparticle stabilizer present in its free state in the nanoparticle solution. As shows both the analysis of literature and our own experience obtained in experiments on cell cultures, the contribution of stabilizer, at least for stabilization with surface active substances (SAS), can be essential. This conclusion issues from the results of control experiments with SAS solutions introduced to the same concentrations as in the synthesis of gold and silver nanoparticles, as well as from the studies on cytotoxicity of SAS solutions, including those used as nanoparticle stabilizers [42–47].

It is worth noting that, as shown in the special investigation [46], the toxic action of free SAS can manifest itself not only if the nanoparticle solution is used directly as obtained after the synthesis, but even if the nanoparticles had been removed from the initial solution, washed 3 times and resuspended in distilled water, i.e., when SAS concentration in the nanoparticle solution does not exceed 100 μmol/L. It is not surprising, since the presence of free SAS in the nanoparticle solution results from the equilibrium between stabilizer molecules in the nanoparticle shell and those in solution. Hence, in principle, it is impossible to maintain the nanoparticles stable in water solution at the free surfactant concentration lower that its equilibrium value at a given temperature. Therefore, in studies of cytotoxicity of the SAS-stabilized nanoparticle solution it is necessary to carry out the control measurements of SAS water solution toxicity. This is justified also for the nanoparticles washed from the initial solution and resuspended in water, but in this case, before the control measurement on stabilizer solution one should determine first the residual stabilizer concentration in thus prepared nanoparticle suspension. Correspondingly, at the elucidation of the mechanism of nanoparticle action on cells the literary data become actual on the toxicity of the relevant surfactants found in experiments with SAS water solutions on microbial or animal cells.

The works in the last direction are carried out with various kinds of SAS, including those applied as nanoparticle stabilizers (SDS, CTAB et al) and with cell types used in studies of the biological effects of nanoparticles. Thus, in studies on the cytotoxicity mechanisms of nanoparticle solutions, apart from the action of only nanoparticles the task appears to elucidate the mechanism of free SAS action on cells. A brief account of our considerations essential for the development of correct methodology in studies of the SAS cytotoxicity is given below.

4.1 *In vitro* studies on the mechanism of SAS action on cells: general remarks

Judging from the literature available, manifestations of SAS cytotoxicity (which underlie their toxic effects towards living organisms) are interpreted as the result of action of surfactant molecules on the structure and functions of cell membrane. To our view, the explanation of cell responses on the level of solely molecular interactions does not exhaust the question about the mechanism of SAS action because, as a rule, in studies of cytotoxicity the initial water SAS solutions are used, with concentrations higher than their critical micelle concentration (CMC), that is, containing both surfactant molecules (monomers) and micelles. These two forms can act by different mechanisms, the contribution of each form being dependent on the object under study and details of experimental design.

That is why in the general case elucidation of the mechanism of SAS effect on cells implies the work in three main directions: (1) separation of monomers' and micelles' contributions and determination of their relation, (2) determination of the monomers' mode of action and (3) determination of the micelles' mode of action. In some cases the cell responses observed may result from the action of only molecules or only micelles; then the study is limited to the two directions. Thus the question whether only monomers or both monomers and micelles are present in a cell medium after the introduction of initial (stock) SAS solution and hence, whether the necessity exists to determine first the separate contribution of each form into the cytotoxicity, cannot be solved otherwise than in the course of experiment allowing to register separately the effects of monomers and micelles.

Taking into account the considerations stated above, as well as the results of our earlier studies on the liposomes' interaction with planar bilayer (as a model of biological vesicle-cell interaction) [48, 49] where it was shown that liposomes and lipid monomers present in liposome suspension manifest different modes of action, it is possible to suggest the version of SAS cytotoxicity mechanism including the two parallel processes: (1) incorporation of SAS molecules into the lipid bilayer (presumably into the external monolayer) of cell membrane and (2) adsorption of SAS micelles on a cell surface. At SAS concentration below its CMC only the monomer incorporation takes place; at its concentration near or higher than CMC both processes are realized.

However, unlike the model system liposomes – planar lipid membrane, for surfactant micelles on a cell surface may take place also the other ways of interaction with cell membrane, depending on the SAS nature, micelle parameters (size, form, surface charge), individual cell properties and experimental conditions. Since at the increase of total SAS concentration above the CMC the monomer concentration remains constant and that of micelles increases, one can suggest that here the increase of toxicity is caused by the increase of micelle contribution into the total toxic effect of surfactant.

The whole picture of interactions between SAS solutions and biological cell described above is confirmed by the experiments fulfilled by us for clearing out the origin of the difference in toxicity of silver nanoparticles stabilized with anionic SAS (AOT) at the concentrations below and above its CMC. Here we present a brief account of the strategy of research and the main results allowing to conclude about the trustworthiness of the toxicity mechanism suggested for the chosen SAS and two cell types and, correspondingly, about the reliability of our experimental approach. A detailed description of the methodology and experimental design may be found in our recent publications [31, 50].

4.2 *In vitro* studies on the toxicity mechanism of anionic SAS (AOT) used as stabilizer of silver nanoparticles

The toxic action of stabilizer has been repeatedly recorded in our studies of the biological effects of AOT-stabilized silver nanoparticles, including the experiments on cultured human cells [11, 51, 52]; for a long time, however, mechanism of the surfactant toxicity remained unclear. It was just several years ago that the favorable conditions formed for the progress in this direction. *First*, the new version of biochemical synthesis was created, allowing the nanoparticle preparation directly in water solution at the low stabilizer concentration [53]. *Second*, basing on this version of synthesis the data were obtained on the cytotoxicity of nanoparticle solutions at AOT concentration around its CMC in water. As a direct impact for the start of our work on the mechanism of AOT toxicity served the results of Inácio et al. [54] where the sharp fall of cell viability had been registered for the anionic surfactant (SDS) at the concentration around or exceeding its CMC (**Figure 7**). These findings gave grounds to suggest that the biological activity of surfactant molecules and micelles may be essentially different (i.e. micelles can be more toxic than monomers) and hence it is possible, in principle, to determine the effect of each form of surfactant present in solution.

Therefore, we have developed a strategy allowing the separation of monomer and micelle contribution using the results of cell viability measurements and estimation of the toxicity changes of each form at the changes of total surfactant concentration in the nanoparticle solution stabilized by this surfactant. This strategy is applicable at the most often used incubation times (24 hours and longer), for the ionic SAS and "native" nanoparticle solutions obtained by the chemical synthesis, which were not subjected to the procedures of nanoparticle separation and resuspending in the other medium (e.g. in water). Below we give the recommended sequence of steps with some illustrations reproduced from our recent article [31] where our approach was applied in studies of the AOT-stabilized AgNPs interaction with malignant Jurkat cells.

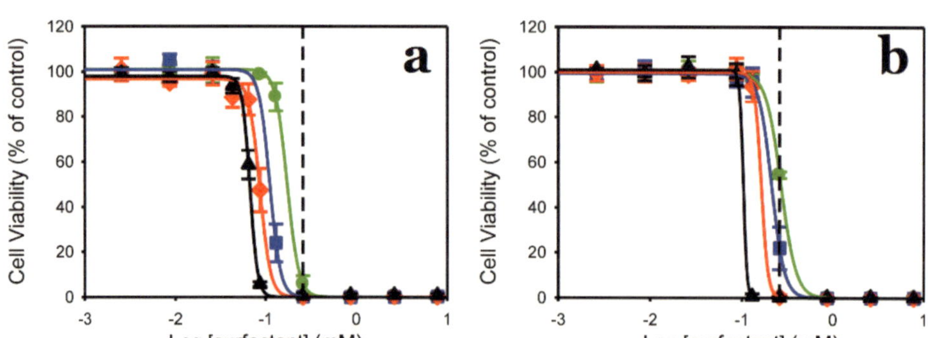

Figure 7.
Effect of SDS on viability of HeLa (a) and FDSC (b) cells. The viability was evaluated using MTT assay 24 h after treatment with SDS for 20 min (green circles), 60 min (blue squares), 180 min (red rhombi), and 540 min (black triangles). Dashed line is the SDS CMC value in serum-free medium. Reproduced from [54].

1. Determination of the CMC value for a stabilizer in distilled water; it may be taken from handbooks (e.g. [55]) or measured using a standard method available (e.g. [56, 57]). In our case the CMC for AOT was 2.81 mM as found from the conductivity measurements.

2. Synthesis of SAS-stabilized nanoparticles by the same method at surfactant concentrations higher and lower than its CMC in distilled water; in our work these concentrations were, respectively, 3 mM (3-AgNPs) and 1 mM (1-AgNPs). Mind that in both solutions the nanoparticle mean size, form and concentration should be the same, and also equal or similar should be the values of zeta potential. These conditions are necessary for the measured toxic effects on cell viability to be caused only by the difference in stabilizer concentration.

3. Measurements of cell viability (V) after the incubation during the same time at the same dilutions of the two named nanoparticle solutions and the respective stock stabilizer solutions. Determination of the difference in viability changes in the same range of dilutions; here the increase of toxicity should be observed for the nanoparticle and stabilizer solutions with concentration exceeding its CMC because of the appearance of surfactant micelles. Indeed, the expected result was observed in our viability measurements (**Figure 8**).

4. To verify the supposed role of surfactant micelles in this increase of nanoparticle toxicity one should find the surfactant CMC values in both solutions. The direct CMC measurement in nanoparticle solutions may be hampered because of the presence of stabilizer in the concentrations near its CMC. It is possible however, to substitute the nanoparticle solutions for the suitable model solutions, so that the CMC can be measured in conditions at most close to those in the nanoparticle solutions with respect to the parameters affecting the CMC value. Since the CMC for the ionic surfactants depends on ionic strength and ionic composition of solution [56, 57], these two parameters should be equal to those which exist in the nanoparticle solution. In our example, considering the measured conductivities and ionic composition of 1-AgNPs and 3-AgNPs, as the most suitable model solutions were used, respectively, 5 mM and 8 mM KNO_3. The relevant CMC values for AOT determined by the conductivity method were found to be, respectively, 2.14 and 1.7 mM (for more details see [31]).

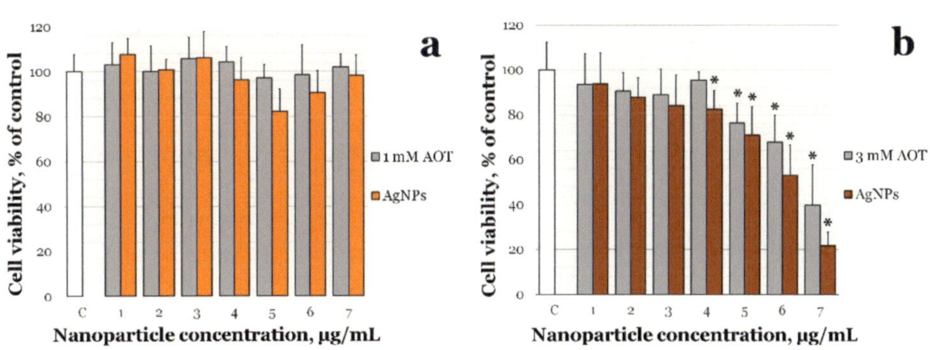

Figure 8.
*Jurkat cell viability dependence on AgNPs concentration. The nanoparticles were synthesized in water solution containing 1 mM AOT (a) and 3 mM AOT (b) and introduced from 108 µg/mL stock AgNPs solution in standard dilutions (108, 54, 36, 27, 21.6, 18, and 15.4) related to AgNPs concentrations in the range 1–7 µg/mL. The corresponding figures are given under the columns. The results obtained with 1 mM and 3 mM AOT water solutions in standard dilutions are also presented. Control – distilled water. * – here and in the* ***Figures 10*** *and* ***11****, the viability values different from control with statistical significance ($p < 0.05$). Reproduced from [31].*

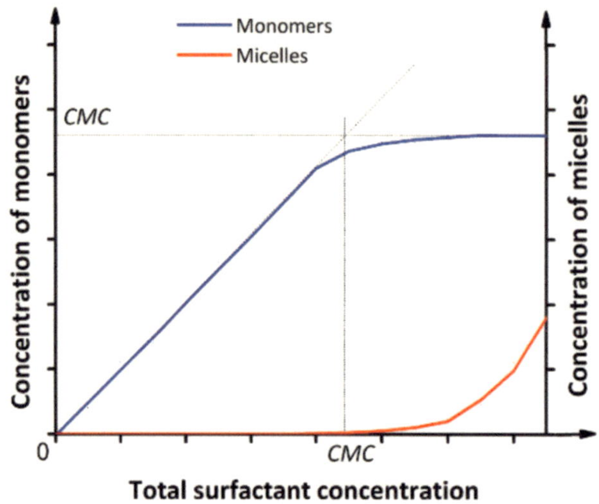

Figure 9.
Changes of monomer and micelle concentrations at low total surfactant concentrations (qualitative presentation). Reproduced from [31].

5. As follows from **Figure 9** which represents the general character of concentration changes for monomers and micelles in the region of small total surfactant concentration (C^{tot} near CMC), the monomer concentration, C^{mon}, is equal to CMC while that of micelles, C^{mic}, at $C^{tot} > CMC$ is equal to the difference:

$$C^{mic} = C^{tot} - C^{mon} = C^{tot} - CMC \tag{8}$$

Relation (8) allows the estimation of micelle concentration in water and (assuming that CMC values in the model solutions are equal to those in the relevant AgNPs) also in the nanoparticle solutions under study. In our research, thus found monomer and micelle concentrations in 1-AgNPs and 3-AgNPs are shown in **Table 4**. As issues from the table, in 1 mM AOT containing solutions (both in distilled water and in 1-AgNPs) the stabilizer exists only in molecular form, while in 3 mM AOT containing solutions this surfactant is present in the two forms – molecules and micelles; however, in its water solution the per cent of micelles is considerably lower than in the 3-AgNPs. Hence follows that, to discover the effect of AOT micelles, in experiment with 3-AgNPs the correct control for the stabilizer toxicity should be carried out not with its water solution, but with the model

System	CMC, mM	Monomers, mM	Micelles, mM	% of total AOT concentration	
				Monomers	Micelles
Water, 1 mM AOT	2.81	1.0	0.0	100.0	0.0
Water, 3 mM AOT	2.81	2.81	0.19	93.7	6.3
AgNPs +1 mM AOT	2.14	1.0	0.0	100.0	0.0
AgNPs +3 mM AOT	1.7	1.7	1.3	56.7	43.3

Table 4.
Measured AOT CMC values, calculated concentrations of AOT monomers and micelles in water and AgNPs solutions, and their contributions to the total AOT concentration.

solution with the same CMC of the surfactant, that is, with the same concentration of its micelles. In our case the correct control was fulfilled with 8 mM KNO_3 solution where the AOT CMC was found to be 1.7 mM, presumably equal to that in 3-AgNPs.

6. Further, the monomer contribution (ΔV^{mon}) can be determined for the nanoparticle and AOT solutions with surfactant concentration higher than CMC. For this purpose, cell viability is measured after the incubation with dilutions of the stock AOT solution in the model salt solution with $C^{tot} = C^{mon} = CMC$, where AOT exists only as monomers. At each dilution (n) the decrease of cell viability is estimated as $\Delta V^{mon} = 100\% - V^n$. In our measurements (**Figure 10**) an example is shown for AOT concentration corresponding to n = 5 in the range of standard dilutions (see legend to **Figure 8**).

7. Then the micelle contribution (ΔV^{mic}) into the toxicity of nanoparticle and SAS solution with surfactant concentration $C^{tot} > CMC$ is determined from the results of the repeated measurements of nanoparticle and SAS toxicity with the use of surfactant dilutions not in water, but in the relevant model salt solution. At each dilution, the summary decrease of cell viability after incubation with SAS (ΔV^{tot}) is obtained, including the monomer and micelle contributions. The micelle contribution is found as $\Delta V^{mic} = \Delta V^{tot} - \Delta V^{mon}$, where ΔV^{mon} is that found as described in p.6. In our example this is illustrated in **Figure 11** for the same dilution as that shown in **Figure 10**. The results obtained for the AOT-stabilized AgNPs on Jurkat cells are summarized in **Table 5.**

8. Finally, the dependences of separate monomer and micelle contributions into the changes of cell viability are obtained for the chosen range of nanoparticle or SAS concentrations. These dependences reflect the difference in the toxic action of these two surfactant forms towards a cell line under study. The corresponding plots created according to the data in **Table 5** are presented in **Figure 12**. It is seen that the micelle contribution into the toxicity of 3 mM

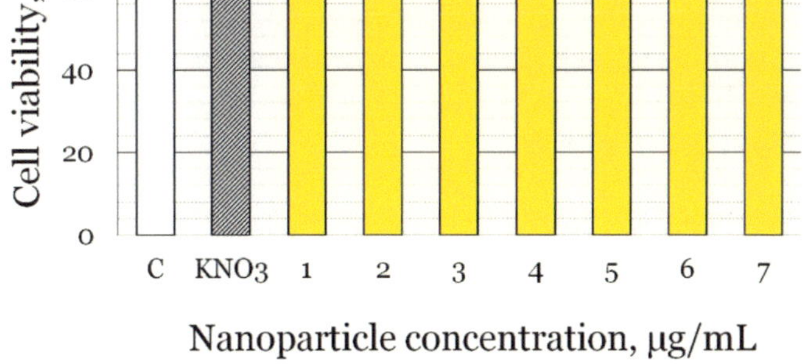

Figure 10.
Viability changes of Jurkat cells treated with AOT monomers. Cells were incubated with 1.7 mM AOT in 8 mM KNO_3 stock solution introduced in the standard dilutions. As an example, cell viability decrease at the dilution corresponding to 5 µg/mL of AgNPs (ΔV_5^{mon}) is shown. Control – distilled water (C) and 8 mM KNO_3 (KN). Adapted from [31].

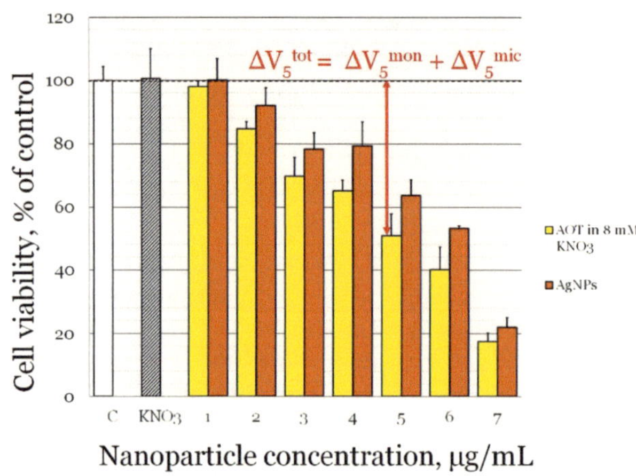

Figure 11.
*Correction of the AOT toxicity in accordance with change of its CMC in the 3-AgNPs solution. Jurkat cells were incubated for 24 h with the standard dilutions of 3-AgNPs and 3 mM AOT in 8 mM KNO_3. At the dilution corresponding to 5 μg/mL of AgNPs, the total AOT contribution to viability decrease (ΔV_5^{tot}) is represented as sum of the corresponding contributions of AOT monomers (ΔV_5^{mon}) and micelles (ΔV_5^{mic}). Controls are the same as in **Figure 10**. Adapted from [31].*

Viability changes	Nanoparticle concentrations (n), μg/mL						
	1	2	3	4	5	6	7
V_n^{mon}	98.5 ± 5.1	92.2 ± 5.1	87.6 ± 5.5	80.6 ± 5.2	78.9 ± 5.5	74.6 ± 5.4	74.0 ± 7.8
ΔV_n^{mon}	1.5	8.8	12.4	19.4	21.1	25.4	26.0
V_n^{tot}	98.0 ± 1.9	84.8 ± 2.1	69.8 ± 5.9	65.2 ± 3.3	50.8 ± 7.2	40.2 ± 7.1	17.4 ± 2.7
ΔV_n^{tot}	2	15.2	30.2	34.8	49.2	59.8	82.6
ΔV_n^{mic}	0.5	6.4	17.8	15.4	28.1	34.4	56.6

*Data taken from **Figures 9** and **10**.*
*Abbreviations: V_n^{mon} and ΔV_n^{mon}, respectively, measured cell viabilities and calculated contributions of monomers at standard dilutions (see **Figure 10**); V_n^{tot} and ΔV_n^{tot}, respectively, measured cell viabilities and calculated changes in viabilities after treatment with 3 mM AOT at standard dilutions (see **Figure 11**); ΔV_n^{mic}, contributions of AOT micelles to cell viability changes calculated from the data of **Figure 11**. Adapted from [31].*

Table 5.
Contributions of monomers and micelles into the total toxic effect of AOT solutions.

AOT solutions (or of the AgNPs with the same stabilizer concentration) is more expressed than that of monomers at the AOT concentrations exceeding those introduced at n = 5. It is clear also that, contrary to the monomer contribution, that of the micelles demonstrates the tendency to exponential growth, the fact which indicates to the different mechanism of action of the two surfactant forms in water solution.

Qualitatively similar changes of monomer and micelle toxicity with changes of the total AOT concentration have been observed on the other cell type – on normal endothelium human cells (EA.hy926 line). In this case also the micelle toxicity changes with surfactant concentration reveal the tendency to exponential growth, while the monomer toxicity demonstrates the smooth increase which passes on plateau. Within the frames of the research strategy suggested, such a result was expected, since this approach supposes a qualitative difference in the mechanism of action between monomers and micelles. At the same time, the absolute values of ΔV^{mic} and ΔV^{mon} at the equal total nanoparticle concentrations differ from those

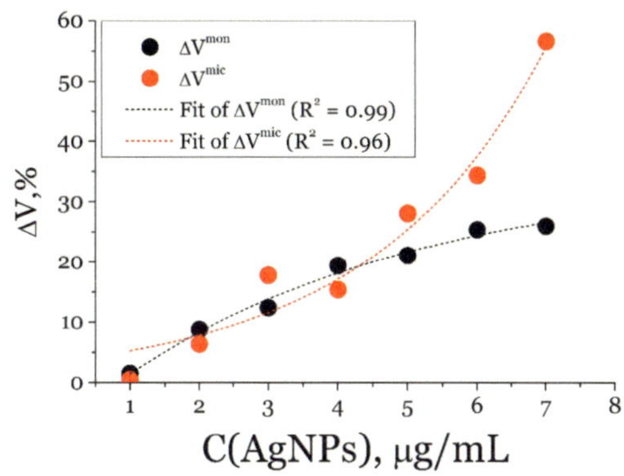

Figure 12.
Contributions of monomers and micelles to the toxic effect of AOT on Jurkat cells at various AgNPs concentrations. The 3-AgNPs solution was used. Abbreviations: ΔV, cell viability changes; ΔV^{mon} and ΔV^{mic}, cell viability changes resulted from the effect of, respectively, AOT monomers and micelles; C(AgNPs), concentration of Ag nanoparticles. Reproduced from [31].

obtained on Jurkat cells; this indicates to the dependence of monomer and micelle action on the individual properties of cells, the fact which may allow to obtain more detailed information about the cytotoxicity mechanism of surfactants.

The approach to studies of the mechanism of SAS cytotoxicity described in this section may be useful for the nanoparticles stabilized with various surfactants. It is important, as we believe, to draw attention to the active role of stabilizer micelles, because this opens the opportunity, *first*, to make the reliable conclusions concerning the mechanism of the nanoparticle action and, *second*, to specify the surfactant concentrations used in experiments on the nanoparticle cytotoxicity as well as in the surfactants' applications in the other fields. If either surfactant solution or that of surfactant-stabilized nanoparticles is introduced into a cell medium by means of dilution of the stock solution with surfactant concentration higher than CMC, it contains both surfactant monomers and micelles and both forms are present in a cell medium after the dilution, in the first moment in the same relation as in the stock solution. Then this relation changes in the process of establishment of the new equilibrium; however, it is unknown how great is the final micelle concentration and how will they affect the measured viabilities or other cell characteristics. Therefore, if it is desirable to determine the toxicity of only molecular surfactant, its concentration in the stock solution must not exceed CMC.

In the last years participation of surfactants in nanoparticle – cell interactions draws attention of many researchers (e.g. [43, 46, 47, 58, 59]). Hence there is hope that the considerations expounded above will favor the progress in the understanding of the mechanism of biological activity of surfactants applied both as nanoparticle stabilizers (in nanotoxicology) and as water solutions in medical practice.

5. Conclusion

In the end we consider it reasonable to emphasize the importance of the three essential points in methodology of studies of the metal nanoparticles' action on biological cells examined in this review.

First, in studies of the particle size effect on the biological activity it is strongly recommended to change the mode of expression of the nanoparticle concentration, namely to pass from the mass to number concentration with the corresponding change of the measured cell parameter dependence on the particle concentration.

As seen from our estimates of the toxic action of differently sized nanoparticles at the same number concentration, accumulation of similar data in the further studies may lead to the revision of the wide-spread opinion about the increase of toxicity with the decrease of nanoparticle size, with the positive consequences for their various applications in medicine.

Second, it is important to make corrections for the relaxation effect in calculations of zeta potentials from the measured electrophoretic mobilities of small particles in water solutions. It seems necessary to understand, as we believe, that there is the problem of invalidity of the Smoluchowski equation for zeta potential determination from the EPM measurements for the small charged particles of different nature, at the definite relations of their size, charge and ionic strength of solution. The error in zeta potential values calculated from the Smoluchowski equation is the more significant, the less is particle size, hence the necessity to take into account the relaxation effect is the most evident in studies of nanoparticles. We believe, however, that the corrections under question are actual not only for the correct estimation of nanoparticle surface charge in a wide variety of experiments, including the elucidation of nanoparticle cytotoxicity, but also probably for the revision of certain estimates of biological effects observed for the other charged particles used in medical and biological researches, e.g. for liposomes, biological vesicles, polymer containers for the delivery of medicines et al. Since, as a rule, the corresponding corrections are not provided by the software installed in devices used for the EPM measurements, we suggest applying for this purpose the equations proposed by the Dukhin theory.

Third, it is possible to change the conventional approach in studies of the mechanism of cytotoxicity both of nanoparticles stabilized with surfactants and of the surfactant solutions used in medical practice for the suppression of various infections. More precisely, the question is about the widening of the range of purposes, because it is suggested to include here the determination of the toxicity effects not only of molecular, but also of micellar form of surfactant, the latter being capable to exert its own contribution to cell responses.

It should be added that, certainly, the recommended changes in methodology do not exhaust the problems faced by researchers in studies of the biological effects of nanoparticles on cultured cells. They only indicate to some actual questions ripe, as we believe, in this direction of nanobiology. The main stimulus for the discussion of the questions raised in this review was our desire to draw attention to these questions and to suggest the ways for their solution, in order to achieve the better, the more reliable results in studies of the biological effects of nanoparticles for the aims of practical medicine.

Conflict of interest

The authors declare no conflict of interest.

Appendices and nomenclature

AgNPs	silver nanoparticles
AOT	aerosol-OT (bis –(2-ethylhexyl) sulphosuccinate, sodium salt)
CMC	critical micelle concentration
CTAB	cethyl trimethyl ammonium bromide
EDL	electrical double layer
EPM	electrophoretic mobility
PCS	photon correlation spectroscopy

SAS	surface active substances
SDS	sodium dodecyl sulphate
TEM	transmission electron microscopy

Author details

Elena Mikhailovna Egorova* and Said Ibragimovich Kaba
Institute of General Pathology and Pathophysiology, Moscow, Russia

*Address all correspondence to: el.magnet60@gmail.com

References

[1] Oberdorster G. Safety assessment for nanotechnology and nanomedicine: concepts of nanotoxicology. Journal of Internal Medicine. 2010;267:89-105. DOI: 10.1111/j.1365-2796.2009.02187.x

[2] Aydın A, Sipahi H, Charehsaz M. Nanoparticles toxicity and their routes of exposures. In: Recent advances in novel drug carrier systems. INTECH; 2012. p. 483-500. DOI: 10.5772/51230

[3] Arora S, Rajwade JM, Paknikar KM. Nanotoxicology and in vitro studies: The need of the hour. Toxicology and Applied Pharmacology. 2012;258:151-165. DOI: 10.1016/j.taap.2011.11.010

[4] Lara HH, Garza-Trevino EN, Ixtepan-Turrent L, Singh DK. Silver nanoparticles as broad-spectrum bactericidal and virucidal compounds. Journal of nanobiotechnology. 2011;9:30. DOI: 10.1186/1477-3155-9-30

[5] Monteiro DR, Gorup LF, Takamiya AS, Ruvollo-Filho AC, de Camargo ER, Barbosa DB. The growing importance of materials that prevent microbial adhesion: antimicrobial effect of medical devices containing silver. International Journal of Antimicrobial Agents. 2009;34:103-110. DOI: 10.1016/j.ijantimicag.2009.01.017

[6] Rai M, Yadav A, Gade A. Silver nanoparticles as a new generation of antimicrobials. Biotechnology Advances. 2009;27:76-83. DOI: 10.1016/j.biotechadv.2008.09.002

[7] Korani M, Ghazizadeh E, Korani S, Hami Z, Mohammadi-Bardbori A. Effects of silver nanoparticles on human health. European Journal of Nanomedicine. 2015;7:51-62. DOI: 10.1515/ejnm-2014-0032

[8] Safari M, Bidgoli SA, Rezayat SM. Differential neurotoxic effects of silver nanoparticles: A review with special emphasis on potential biomarkers. Nanomedicine Journal. 2016;3:83-94. DOI: 10.7508/nmj.2016.02.002

[9] Zhang X-F, Shen W, Gurunathan S. Silver nanoparticle-mediated cellular responses in various cell lines: an in vitro model. International Journal of Molecular Sciences. 2016;17:1603; DOI:10.3390/ijms17101603

[10] Egorova EM, Kaba SI, Kubatiev AA. Toxicity of silver nanoparticles obtained by bioreduction as studied on malignant cells: is it possible to create new generation of anticancer remedies? In: Grumezescu AM, editor. Nanobiomaterials in Cancer Therapy. 1st ed. Amsterdam: Elsevier; 2016. p.505-542. DOI: 10.1016/B978-0-323-42863-7.00015-3

[11] Egorova EM, Kubatiev AA, Schvets VI. Biological Effects of Metal Nanoparticles. Springer International Publishing; 2016. 292 p. DOI: 10.1007/978-3-319-30906-4

[12] Egorova EM, Kubatiev AA. Nanotechnology: methodology of the studies of metal nanoparticle action on biological objects. Moscow: Urait; 2020. 188 p. (in Russian)

[13] Henry DC. The electrophoresis of suspended particles. IV. The surface conductivity effect. Transactions of the Faraday Society. 1948;44:1021-1030. DOI: 10.1039/TF9484401021

[14] Booth F. Surface conductance and cataphoresis. Transactions of the Faraday Society. 1948;33:955-963. DOI: 10.1039/TF9484400955

[15] Wiersema PH, Loeb A, Overbeek JTG. Calculation of the electrophoretic mobility of a spherical colloid particle. Journal of Colloid and Interface Science. 1966;22:78-99.

[16] Dukhin SS, Deryaguin BV. Electrokinetic Phenomena, in: Matijević E, editor. Surface and Colloid Sciences, Vol. 7. New York: Wiley; 1974. p. 335

[17] Dukhin SS, Deryaguin BV. Electrophoresis. Moscow: Nauka; 1976. 328 p. (In Russian)

[18] O'Brien RW, White LRJ. Electrophoretic mobility of a spherical colloidal particle. Journal of the Chemical Society, Faraday Transactions 2: Molecular and Chemical Physics. 1978;74:1607-1626. DOI: 10.1039/F29787401607

[19] O'Brien RWJ. The solution of the electrokinetic equations for colloidal particles with thin double layers. Journal of Colloid and Interface Science. 1983;92: 204-216. DOI: 10.1016/0021-9797(83)90129-7

[20] Hunter RJ. Zeta Potential in Colloid Science. 1st ed. New York: Academic Press: 1981. 398 p. DOI: 10.1016/C2013-0-07389-6

[21] Belenkii D, Balakhanov D, Lesnikov E. Measurement of the zeta potential. Brief review of the main methods. Analytics. 2017;3:82-89. DOI: 10.22184/2227-572X.2017.34.3.82.89 (In Russian)

[22] Panáček A, Kvítek L, Prucek R, Kolář M, Večeřová R, Pizúrová N, Sharma VK, Nevěčná T, Zbořil R. Silver colloid nanoparticles: Synthesis, characterization, and their antibacterial activity. Journal of Physical Chemistry B. 2006;110:16248-16253. DOI: 10.1021/jp063826h

[23] Kim T-H, Kim M, Park H-S, Shin US, Gong MS, Kim HW. Size-dependent cellular toxicity of silver nanoparticles. Journal of Biomedical Materials Research - Part A. 2012:100A:1033-1043. DOI: 10.1002/jbm.a.34053

[24] Prasad RY, McGee JK, Killius MG, Suarez DA, Blackman CF, DeMarini DM, Simmons SO. Investigating oxidative stress and inflammatory responses elicited by silver nanoparticles using high-throughput reporter genes in HepG2 cells: Effect of size, surface coating, and intracellular uptake. Toxicology in vitro. 2013;27:2013-2021. DOI: 10.1016/j.tiv.2013.07.005

[25] Butler KS, Peeler DJ, Cassey BJ, Dair BJ, Elespuru RK. Silver nanoparticles: correlating nanoparticle size and cellular uptake with genotoxicity. Mutagenesis. 2015;30:577-591. DOI:10.1093/mutage/gev020

[26] Ivask A, Visnapuu M, Vallotton P, Marzouk ER, Lombi E, Voelcker NH. Quantitative multimodal analyses of silver nanoparticle-cell interactions: Implications for cytotoxicity. NanoImpact. 2016;1:29-38. DOI: 10.1016/j.impact.2016.02.003

[27] Guo X, Li Y, Yan J, Ingle T, Jones MY, Mei N, Boudreau MD, Cunningham CK, Abbas M, Paredes AM, Zhou T, Moore MM, Howard PC, Chen T. Size- and coating-dependent cytotoxicity and genotoxicity of silver nanoparticles evaluated using in vitro standard assays. Nanotoxicology. 2016;10:1373-1384. DOI: 10.1080/17435390.2016.1214764

[28] Haase A, Tentschert J, Jungnickel H, Graf P, Mantion A, Draude F, Plendl J, Goetz ME, Galla S, Mašić A, Thuenemann AF, Taubert A, Arlinghaus HF, Luch A. Toxicity of silver nanoparticles in human macrophages: uptake, intracellular distribution and cellular responses. Journal of Physics: Conference Series. 2011;304:012030. DOI: 10.1088/1742-6596/304/1/012030

[29] Makama S, Kloet SK, Piella J, van den Berg H, de Ruijter NCA, Puntes VF, Rietjens IMCM, van den Brink NW. Effects of systematic variation in size and surface coating of silver nanoparticles on their in vitro toxicity

to macrophage RAW 264.7 cells. Toxicological Sciences. 2018;162:79-88. DOI: 10.1093/toxsci/kfx228

[30] Perde-Schrepler M, Florea A, Brie I, Virag P, Fischer-Fodor E, Vâlcan A, Gurzău E, Lisencu C, Maniu A. Size-dependent cytotoxicity and genotoxicity of silver nanoparticles in cochlear cells in vitro. Journal of Nanomaterials. 2019:6090259. DOI: 10.1155/2019/6090259

[31] Egorova EM, Kaba SI. The effect of surfactant micellization on the cytotoxicity of silver nanoparticles stabilized with aerosol-OT. Toxicology in vitro. 2019;57:244-254. DOI: 10.1016/j.tiv.2019.03.006

[32] McLaughlin S. The electrostatic properties of membranes. Annual Review of Biophysics and Biophysical Chemistry. 1989;18:113-136. DOI: 10.1146/annurev.bb.18.060189.000553

[33] Cevc G. Membrane electrostatics. Biochimica et Biophysica Acta (BBA) - Reviews on Biomembranes. 1990;1031:311-382. DOI: 10.1016/0304-4157(90)90015-5

[34] Kumal RR, Karam TE, Haber LH. Determination of the surface charge density of colloidal gold nanoparticles using second harmonic correlation. Journal of Physical Chemistry C. 2015;119:16200-16207. DOI: 10.1021/acs.jpcc.5b00568

[35] Egorova EM, Dukhin AS, Svetlova IE. Some problems of zeta potential determination in electrophoretic studies on lipid membranes. Biochimica et Biophysica Acta (BBA) - Biomembranes. 1992;1104:102-110. DOI: 10.1016/0005-2736(92)90137-B

[36] Egorova EM, Yakover LL, Svitova TF. The relaxation effect as observed on lipid suspensions of low polydispersity. Biochimica et Biophysica Acta (BBA) - Biomembranes. 1992; 1109:1-6. DOI: 10.1016/0005-2736(92)90180-T

[37] Egorova EM. The validity of the Smoluchowski equation in electrophoretic studies of lipid membranes. Electrophoresis. 1994;15:1125-1131. DOI: 10.1002/elps.11501501170

[38] Midmore BR, Hunter RW. The effect of electrolyte and co-ion type on the ζ-potential of polystyrene latexes. Journal of Colloid and Interface Science. 1988;122:521-529. DOI: 10.1016/0021-9797(88)90387-6.

[39] Egorova EM. Disagreement between experimental potential changes and the Gouy-Chapman theory predictions for dilute 1 : 1 electrolyte solutions : 1. Phenomenology of a problem. Colloid Journal. 2000;62:284-296.

[40] Ulberg ZR, Dukhin AS. Electrodiffusiophoresis - film formation in Ac and Dc electrical fields and its application for bactericidal coatings. Progress in Organic Coatings. 1990;18:1-41. DOI: 10.1016/0033-0655(90)85001-E

[41] Egorova EM. Some applications of the Dukhin theory in studies of lipid membranes. Colloids and Surfaces A: Physicochemical and Engineering Aspects. 2001;192:317-330. DOI: 10.1016/S0927-7757(01)00732-4

[42] Yang L, Watts DJ. Particle surface characteristics may play an important role in phytotoxicity of alumina nanoparticles. Toxicology Letters. 2005:122-132. DOI 10.1016/j.toxlet.2005.03.003

[43] Wang S, Lu W, Tovmachenko O, Rai US, Yu H, Ray PC. Challenge in understanding size and shape dependent toxicity of gold nanomaterials in human skin keratinocytes. Chemical Physics Letters.

2008;463:145-149. DOI: 10.1016/j.cplett.2008.08.039

[44] Lu W, Senapati D, Wang S, Tovmachenko O, Singh AK, Yu H, et al. Effect of Surface Coating on the Toxicity of Silver Nanomaterials on Human Skin Keratinocytes. Chemical Physics Letters. 2013;18:1199-1216. DOI: 10.1016/j.cplett.2010.01.027

[45] Ying E, Hwang H-M. In vitro evaluation of the cytotoxicity of iron oxide nanoparticles with different coatings and different sizes in A3 human T lymphocytes. Science of The Total Environment. 2010;408:4475-4481. DOI: 10.1016/j.scitotenv.2010.07.025

[46] Zhang Y, Newton B, Lewis E, Fu PP, Kafoury R, Ray PC, Yu H. Cytotoxicity of organic surface coating agents used for nanoparticles synthesis and stability. Toxicology in vitro. 2015;29:762-768. Available from: DOI: 10.1016/j.tiv.2015.01.017

[47] Zhang Y, Li X, Yu H. Toxicity of nanoparticle surface coating agents: Structure-cytotoxicity relationship. Journal of Environmental Science and Health – Part C Environmental Carcinogenesis and Ecotoxicology Reviews. 2016;34:204-215. DOI: 10.1080/10590501.2016.1202762

[48] Egorova EM. Interaction of liposomes with planar bilayer: molecular diffusion. Electrokhimia. 1984;20:1396-1399. (in Russian)

[49] Egorova EM. Interaction of liposomes with planar bilayer: adsorption. Electrokhimia. 1984;20:1527-1531. (in Russian)

[50] Egorova EM, Kaba SI. Correction of the toxicity control for stabilizer in studies of the silver nanoparticle effect on Jurkat cells. Patologicheskaya Fiziologiya i Eksperimental'naya Terapiya (Pathological physiology and experimental therapy. 2018;62:269-272. DOI: 10.25557/0031-2991.2018.04.269-272

[51] Kaba, S.I., Egorova, E.M., 2015. In vitro studies of the toxic effects of silver nanoparticles on HeLa and U937 cells. Nanotechnol. Sci. Appl. 8, 19-29. doi:10.2147/NSA.S78134.

[52] Egorova, E.M., Kaba, S.I., Tlupova, S.A., 2016. Assessment of the Cytotoxicity of Silver Nanoparticles with Different Surface Charge, in: Bartul, Z., Trenor, J. (Eds.), Advances in Nanotechnology. Nova Science Publishers, New York, pp. 25-48.

[53] Egorova, E.M., Sosenkova, L.S., inventors; Egorova EM, assignee. Russian Federation patent RU 26182270 C1. 2016 Jun 1.

[54] Inácio ÂS, Mesquita KA, Baptista M, Ramalho-Santos J, Vaz WLC, Vieira O V. In vitro surfactant structure-toxicity relationships: Implications for surfactant use in sexually transmitted infection prophylaxis and contraception. PLoS One. 2011;6:e19850. DOI: 10.1371/journal.pone.0019850

[55] Mukherdjee, P. and K. Mysels, K. Critical Micelle Concentration of Various Surfactant Systems, Washington, 1971.

[56] Tanford C. The hydrophobic effect. Formation of micelles and biological membranes. 2nd ed. New York: Wiley; 1980. 233 p.

[57] Rosen MJ. Surfactants and interfacial phenomena. 3rd ed. New York: Wiley; 2004. 444 p. DOI: 10.1002/0471670561

[58] Vinković Vrček I, Pavičić I, Crnković T, Jurašin D, Babič M, Horák D, Lovrić M, Ferhatović L, Ćurlin M, Gajović S. Does surface coating of metallic nanoparticles modulate their interference with in vitro

assays? RSC Advances. 2015;5:70787-70807. DOI: 10.1039/C5RA14100A

[59] Gossmann R, Langer K, Mulac D. New perspective in the formulation and characterization of didodecyldimethylammonium bromide (DMAB) stabilized poly(lactic-co-glycolic acid) (PLGA) nanoparticles. PLoS One. 2015;10:1-15. DOI: 10.1371/journal.pone.0127532

Chapter 5

Cytotoxicity Study

Bhavin R. Chavda and Bhavesh N. Socha

Abstract

The nature of being poisonous to cells is cytotoxicity. A number of cell fates can result in the treatment of cells with a cytotoxic compound. In this study, the nauplii were exposed to different concentrations of compounds for 24 hours. Experiments were concluded with control and different concentration of the compound. The whole set was triplicate to get an accurate result. The LC50 was determined from the best fit line, of a graph of % mortality vs. concentration of nauplii. The LC50 value was obtained from the antilogarithm of log [complex] at 50% mortality (LC50), all data were collected from three independent experiments.

Keywords: cytotoxicity study, Cytotoxicity, Ag-SMX

1. Introduction

The nature of being poisonous to cells is cytotoxicity [1–3]. In evaluating the possible toxicity of a research material, whether plant extracts or biologically active substances derived from plants, cytotoxicity tests are a helpful tool for initial step. For the effective production of a pharmaceutical or cosmetic preparation, minimal to no toxicity is necessary and cellular toxicity studies play a crucial role in this respect. While contemplating the interaction between acute toxicity and cytotoxicity, the principle of basal cytotoxicity, where deleterious effects are noted on structures and functions common to all human cells, is important [4–6]. To classify substances with promising biological activity and marginal cytotoxicity, the selectivity index is a significant test. In order to determine the cytotoxicity of African medicinal plants, various bioassays and a variety of different cell lines have been used. In addition, the extraction of solvents differing in polarity has been used to remove various parts of plants, adding to the wide range of African plants cytotoxicity findings [7–10].

A number of cell fates can result in the treatment of cells with a cytotoxic compound. The cells can undergo necrosis in which as a result of cell lysis, they lose membrane integrity and die quickly. The cells can stop growing and dividing aggressively (a decline in cell viability), or the cells can trigger a regulated cell death genetic programme (apoptosis).

Usually, cells undergoing necrosis show accelerated swelling, lose the cohesion of the membrane, shut down metabolism, and release their contents into the atmosphere. Cells undergoing rapid in vitro necrosis do not have enough time or resources to activate the apoptotic machinery and do not produce apoptotic markers [11–13]. Apoptosis is distinguished by well-defined cytological and molecular events, including cell refractive index transition, cytoplasmic shrinkage, nuclear condensation, and regular fragment cleavage of DNA. Cells undergoing apoptosis in culture ultimately undergo secondary necrosis. The metabolism will be shut down, membrane integrity and lysis will be destroyed.

In compound libraries, cytotoxicity assays are commonly used by the pharmaceutical industry to test for cytotoxicity. If they are interested in creating a therapeutic that targets, for example, rapidly dividing cancer cells, researchers can either look for cytotoxic compounds; or they can scan 'hits' from initial high-throughput drug screens for unintended cytotoxic effects before investing in their development as a pharmaceutical.

2. Examples

Examples of toxic agents are an immune cell or some types of venom, e.g. from the puff adder (*Bitis arietans*) or brown recluse spider (*Loxosceles reclusa*).

3. Application

The prediction of cytotoxicity of chemical compounds based on prior measurements, i.e. in-silico research, is a highly important topic. Several QSAR and virtual screening methods have been suggested for this purpose. In the "Toxicology in the 21st century" project, an objective comparison of these approaches has been carried out.

Chemotherapy as a cancer treatment also relies on the ability of cytotoxic agents to kill or destroy reproductive cells, preferably targeting the rapid division of cancer cells.

Antibody-dependent cell-mediated cytotoxicity (ADCC) explains the ability of certain lymphocytes to kill cells, and involves an antibody to recognise the target cell. On the other hand, cytotoxicity mediated by lymphocytes does not need to be mediated by antibodies; neither does complement-dependent cytotoxicity (CDC) mediated by the complement system.

4. Measurement

One of the most common approaches to test cell viability and cytotoxic effects is to determine cell membrane integrity. Compounds which have cytotoxic effects frequently undermine the integrity of cell membranes. Vital dyes such as trypan blue or propidium iodide are typically removed from the inside of healthy cells; however, they easily cross the membrane and stain components which is intracellular.

Alternatively, by controlling the movement of molecules that are usually sequestered within cells to the outside, membrane integrity can be measured. Using LDH research, one enzyme, lactate dehydrogenase (LDH), is normally tested. LDH reduces NAD to NADH, which, by contact with a particular probe, induces a colour transition.

Protease biomarkers have been established that enable researchers inside the same cell population to quantify relative numbers of live and dead cells. Live cell protease is involved only in cells with a stable cell membrane and, if the cell is infected and the protease is exposed to the external environment, it loses activity. After cells have lost their membrane integrity, the dead-cell protease can not cross the cell membrane, and can only be tested in culture media.

It is also possible to track cytotoxicity using 3-(4, 5-Dimethyl-2-thiazolyl)-2, 5-diphenyl-2H-tetrazolium bromide (MTT) or 2,3-bis(2-methoxy-4-nitro-5-sulphophenyl)-2H-tetrazolium-5-carboxanilide (XTT) developed by a water-soluble substance or MTS assay. Using a colorimetric reaction, this assay tests the decreasing potential of the cell. The MTS reagent would be reduced to a coloured formazan substance by viable cells. Using the fluorescent dye, resazurin, a related redox-based assay was also produced. In addition to using dyes to display cells' redox potential to track their viability, researchers have developed assays that use the substance of ATP as a viability marker. These ATP-based assays provide bioluminescent assays in which ATP is the luciferase reaction's limiting reagent.

Cytotoxicity can also be measured by the sulforhodamine B (SRB) assay, WST assay and clonogenic assay. Suitable assays can be combined and performed sequentially on the same cells in order to reduce assay-specific false positive or false negative results. A possible combination is LDH-XTT-NR (Neutral red assay)-SRB which is also available in a kit format.

A label-free approach to follow the cytotoxic response of adherent animal cells in real-time is based on electric impedance measurements when the cells are grown on gold-film electrodes. This technology is referred to as electric cell-substrate impedance sensing (ECIS). Label-free real-time techniques provide the kinetics of the cytotoxic response rather than just a snapshot like many colorimetric end-point assays.

1. A highly important topic is the prediction of cytotoxicity of chemical compounds based on previous measurements, i.e. in-silico testing [5]. For this purpose many QSAR and virtual screening methods have been suggested. An independent comparison of these methods has been done within the "Toxicology in the 21st century" project.

2. Chemotherapy as a treatment of cancer often relies on the ability of cytotoxic agents to kill or damage cells which are reproducing; this preferentially targets rapidly dividing cancer cells.

3. Antibody-dependent cell-mediated cytotoxicity (ADCC) describes the cell-killing ability of certain lymphocytes, which requires the target cell being marked by an antibody. Lymphocyte-mediated cytotoxicity, on the other hand, does not have to be mediated by antibodies; nor does complement-dependent cytotoxicity (CDC), which is mediated by the complement system.

5. Experimental details

In this study, the nauplii were exposed to different concentrations of compounds for 24 hours. Experiments were concluded with control and different concentration of the compound. The whole set was triplicate to get an accurate result. The LC50 was determined from the best fit line, of a graph of % mortality vs. concentration of nauplii. The LC50 value was obtained from the antilogarithm of log [complex] at 50% mortality (LC50), all data were collected from three independent experiments (**Figure 1**).

6. Materials and physical measurements

Sulfamethoxazole (SMX) sulfa derivatives and d-block transition metal Ag salts are used from Sigma Aldrich and Alfa Aesar respectively, all other reagents are of the highest grade, commercially available and used without further purification. The cellular level in vivo cytotoxic analysis of the free SMX molecule and its Ag -SMX complex were carried out using eukaryotic *S. pombe* (*Schizosaccharomyces pombe*) cell via trypan blue assay.

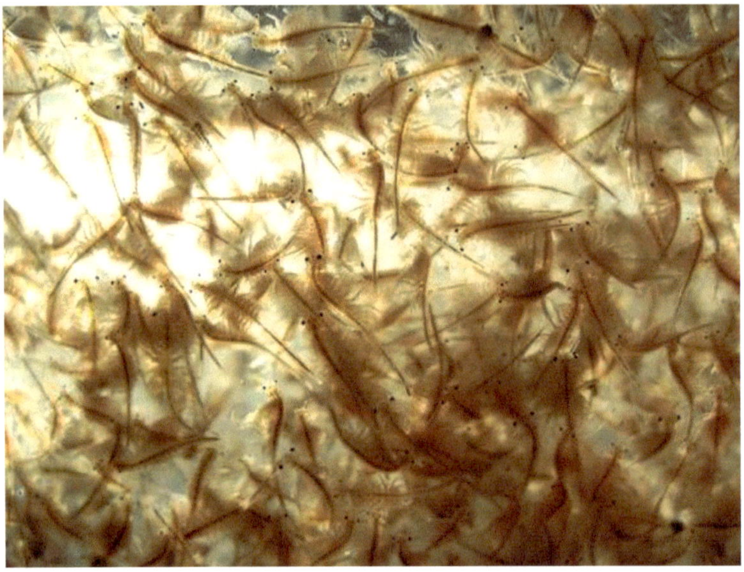

Figure 1.
125 ml live baby brine shrimp coral fish [14].

7. Synthesis of the complex

The silver complex of SMX was synthesized by sulfamethoxazole (SMX) ($C_{10}H_{10}N_3O_3S$, Sigma Aldrich) (0.5 gm, 2 mmol) dissolving in 15 mL methanolic solution (2 mmol, pH 8–9), the metal solution of silver nitrate hydrate ($AgNO_3$, Alfa Aesar) (0.37 gm, 2 mmol) dissolved in deionized water and added drop wise in SMX solution under reflux at 25°C temperature. After stirring for 3 hours, the resultant white colored precipitate of silver compound filtered and washed with mixture of methanol and deionized water (10 ml) respectively, dried in a desiccator. The anhydrous yield of the complex was around 70%. The CHN analysis results for $Ag[C_{10}H_{10}N_3O_3S_1]_2$ are: Anal. Cal. (%): C, 33.35; H, 2.78; N, 11.66. Found (%): C, 33.39; H, 2.94; N, 11.81. pure synthesized compound dissolved in β-picoline solvent and kept for recrystallization at 30°C constant temperature. After almost three months, colorless transparent diamond shaped single crystals were possible to grow (**Figure 2**).

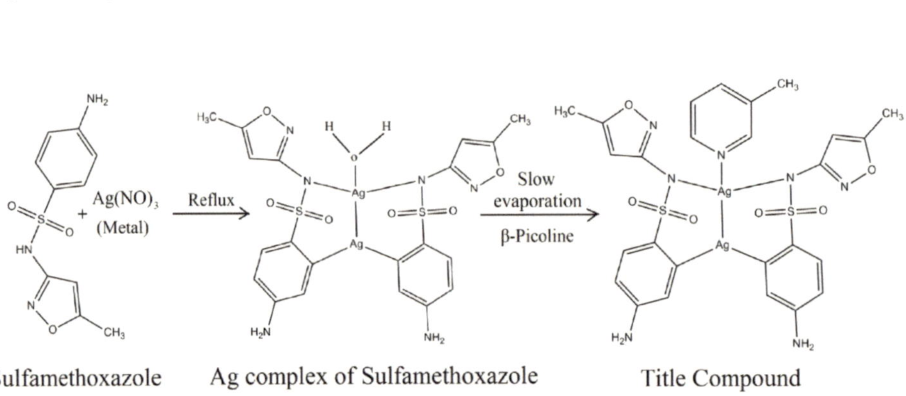

Figure 2.
Steps of reaction proposed.

8. Result and discussion

Brine shrimp lethality assay test protocol was carried out with control and different concentration of the compound. The whole set was triplicate to get an accurate result. The LC50 was determined from the best fit line, of a graph of % mortality vs. concentration of nauplii. The LC50 value was obtained from the antilogarithm of log [complex] at 50% mortality (LC50), all data were collected from three independent experiments. The LC50 (50% lethal concentration) value of the complex had been obtained from the plot of % of the brine shrimp nauplii killed against the complex concentrations and the best-fit line was found from the data through regression analysis. The significant LC50 value of the free Sulfamithoxazole (SMX) drug and its Ag-SMX complex are 12.88 μM and 6.30 μM respectively, predicting higher toxicity of SMX molecule than that of Ag-SMX complex and also, Ag-SMX complex displays less toxic behaviour compared to reported data of Cd sulfapyridine (8.32 μM) and Zn-sulfapyridine (6.31 μM) complexes [12].

The cellular level in vivo cytotoxic analysis of the free SMX molecule and its Ag -SMX complex were carried out using eukaryotic *S. pombe* (*Schizosaccharomyces pombe*) cell via trypan blue assay. **Figure 3** is the plot of % viability of the cells treated with a series of various concentrations 2, 4, 6, 8 and 10 μg/mL of free SMX molecule and its Ag-SMX complex along with control after 17 hours of treatment.

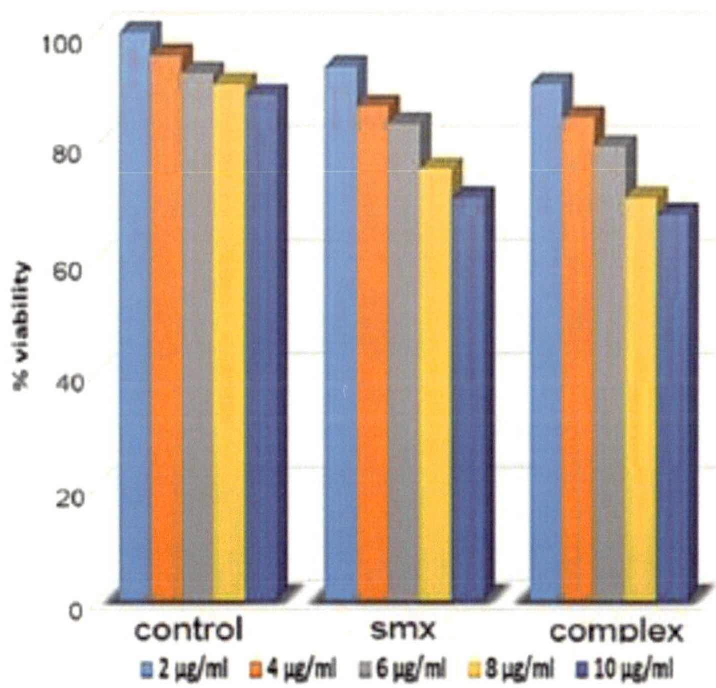

Figure 3.
The plot of % viability of the cells treated with a series of concentrations 2, 4, 6, 8 and 10 μg/mL of SMX molecule and its silver complex respectively along with control after 17 hours of treatment.

9. Conclusion

The significant LC50 value of the free Sulfamithoxazole (SMX) drug and its Ag-SMX complex are 12.88 μM and 6.30 μM respectively, predicting higher toxicity of SMX molecule than that of Ag-SMX complex and also, Ag-SMX complex displays less toxic behaviour compared to reported data of Cd sulfapyridine (8.32 μM) and Zn-sulfapyridine (6.31 μM) complexes.

Author details

Bhavin R. Chavda* and Bhavesh N. Socha
X-ray Laboratory, Department of Physics, Sardar Patel University, Gujarat, India

*Address all correspondence to: chavdabhavin9@gmail.com

References

[1] Lewinski, Nastassja, Vicki Colvin, and Rebekah Drezek. "Cytotoxicity of nanoparticles." small 4.1 (2008): 26-49.

[2] Korzeniewski, Carol, and Denis M. Callewaert. "An enzyme-release assay for natural cytotoxicity." Journal of immunological methods 64.3 (1983): 313-320.

[3] Russell, John H., and Timothy J. Ley. "Lymphocyte-mediated cytotoxicity." Annual review of immunology 20.1 (2002): 323-370.

[4] Banati, Richard B., et al. "Cytotoxicity of microglia." Glia 7.1 (1993): 111-118.

[5] Amos, D. B., et al. "A simple micro cytotoxicity test." Transplantation 7.3 (1969): 220-223.

[6] Chavda, Bhavin R., et al. "Coordination behavior of dinuclear silver complex of sulfamethoxazole with solvent molecule having static rotational disorder: Spectroscopic characterization, crystal structure, Hirshfeld surface and antimicrobial activity." Journal of Molecular Structure (2020): 129777.

[7] Chaudhary, Kaushik P., et al. "Coordination behavior of succinylsulfathiazole–Crystal structure of [Cu (SST).(Pyridine) 3. H2O] n, DNA interaction and cytotoxic studies." Journal of Molecular Structure 1225: 129262.

[8] Chaudhary, Kaushik P., et al. "Divalent zinc complex of succinylsulfathiazole: Synthesis, spectral, molecular structure, DNA interaction profile and Hirshfeld surface analysis." Inorganic Chemistry Communications 122 (2020): 108318.

[9] Socha, Bhavesh N., et al. "Gas-phase DFT studies, quantum chemical calculation and 3D energy framework analysis of novel copper complex of sulfadimethoxine in the presence of secondary ligand 3-methyl pyridine." Inorganic and Nano-Metal Chemistry (2020): 1-9.

[10] Dubey, Rahul P., et al. "Quantum chemical calculations, molecular docking, Hirshfeld surface analysis, biological activity, and characterizations (IR and NMR) of cadmium complex of triple sulfa drugs constituent sulfamethazine." Inorganic and Nano-Metal Chemistry (2020): 1-10.

[11] Pandya, Sachin B., et al. "Combined experimental and computational studies on molecular structure of nickel complex of 4-amino-N-(1, 3-thiazol-2-yl) benzenesulfonamide with coordinated pyridine." Inorganic and Nano-Metal Chemistry (2020): 1-11.

[12] Socha, Bhavesh N., et al. "Role of metal Cu (II) and Ag (I) on molecular structure of 4-amino-N-(2, 6-dimethoxy-pyrimidin-4-yl) benzenesulfonamide in presence of 3-methyl pyridine: Synthesis, spectral, crystallographic and DNA interaction studies." Polyhedron 188 (2020): 114696.

[13] Chavda, Bhavin R., et al. "A quantitative analysis of weak inter-molecular interactions & quantum chemical calculations (DFT) of novel chalcone derivatives." AIP Conference Proceedings. Vol. 1728. No. 1. AIP Publishing LLC, 2016.

[14] https://www.algaebarn.com/blog/live-foods/brine-shrimp/a-beginners-guide-to-brine-shrimp/

Chapter 6

Evolvulus alsinoides (Linn.) Linn.: A Revitalizer

Harshavardhini Kuppusamy R, Jananipriya Srinivasan, Pavithra Sundaramoorthy and Kannan Kilavan Packiam

Abstract

Herbal medicines are utilized as traditional and alternative therapy to precisely restore declining metabolic functions. Herbal bioactive claims its rewards for their effectiveness, safety, and acceptability. *Evolvulus alsinoides* (Linn.) Linn. is a Virya herb from Convolvulaceae with Tridoshhara, Majjadhatu Rasayan, Vajikarak, Chedan, and Nidrajanan properties. In Ayurvedic medicine, the whole plant is used in the treatment of neurodegenerative diseases as brain tonic, amnesia and asthma, epilepsy and as a hepatoprotective. The phytochemical analysis has reported the existence of biomolecules such as β-sitosterol, scopolin, scopoletin, umbelliferon, triacontane, shankpushpine and betaine. A thorough review of classical as well as contemporary literature study has been done on *Evolvulus alsinoides* (Linn.) Linn. to validate the pharmacological actions and cytotoxicity effects on various cell lines.

Keywords: *Evolvulus alsinoides* (Linn.) Linn., virya herb, phytochemical analysis, pharmacological actions, cytotoxicity effects

1. Introduction

The medical herbs are advantageous for humans for the reason of providing remedies that ease the human suffering. The indigenous system has innumerable plants that have been disclosed to have action opposing CNS disorders thus elevating human hardships [1]. The potential of Ayurveda herbs benefit medicinally and it seems likely exploring that medicinal potential is increasing latterly. At the current times, plant dependant drugs procure profuse attention due to the fact that there are onsets of newer approaches on chemical characterization and pharmacological inquisitions [1]. For a long period of time medicinal plants have been valued for their healing and diminishing pain. Medicinal plants are also have been relied abundantly for their curative properties. Various types of herbs have been employed in prominent folk medicine and they have a long history for being benefited in the traditional remedies [2]. In vitro screening methods also contribute to the demand on vital primary inquisition which is crucial to retrieve desired plant extract with promising and effective attributes for upcoming chemical and pharmacological research.

2. *Evolvulus alsinoides* (Linn.) Linn.

The traditional methodology utilized variety of medicinal herbs and one of the notable herbs is *Shankhpushpi* also known as *Evolvulus alsinoides* (Linn.) Linn., has been employed in clinical use for centuries. *Shankhpushpi* is contemplated as *Medha Rasayana* attributes to a drug that revivifies, retains and potentiates intellect and cognizance [3]. *Evolvulus alsinoides* (Linn.) Linn. is included in the Convolvulaceae family and it is a weed primarily available in tropical and sub-tropical miry zones of the world, particularly in East Asia. They are commonly seen in the regions including India and West Cameroon. They are expansively scattered in tropical zones of Africa and world.

Evolvulus alsinoides (Linn.) Linn. is a supine perennial herb consisting small branched rootstock of wood. They are enclosed with plentiful branches that are annual and beyond 30 cm long. The branches include long hairs and frequently prostrate. *Evolvulus alsinoides* (Linn.) Linn. incorporates small leaves which are elliptic in shape. They are acute, susceptive and are densely hairy [4].

Each part of this plant is put to use in Ayurveda medicine to treat cough, cold and fever. They are also employed to cure neurodegenerative diseases like adenitis and dementia. Treatment of venereal disease also includes the use of this plant [5]. Azoospermia, nootroppic and anti-inflammatory activity has also been noted. This plant not only possesses these effects it also consists anti-haemorrhagic and antioxidant effects. Ancient medicine claimed this plant as brain tonic and in current years it has been justifies through pre-clinical research [6]. This plant is preferred in the treatment of asthma and amnesia, most common neurodegenerative diseases. Most notable property of *E.alsinoides* (L.) L. is to boost the memory and elevate the intellectuality [7].

3. Phytochemicals

In current years, Phytochemistry or plant chemistry has been established into evident discipline which lies between organic chemistry of natural creations and biochemistry dealing with plants. It is closely related to both of the subjects [8]. Phytochemistry is related to the innumerable distinct amount of organic materials that are concerned with and aggregated by plants and it deals with plant's properties. The chemical arrangements, biosynthesis, metabolisms and turn over, biological function and their natural distribution.

Evolvulus alsinoides (Linn.) Linn. was interpreted to find the phytochemicals like alkaloids, carbohydrates, steroidal glucosides, saponin, tannins, pseudo tannins, chlorogenic acids, flavones, flavonoids, coumarin, anthocyanin, phenol, terpenoides, resins, volatile oil, anthraquinones, phytosterol and triterpenoids. The analysis portrayed the potentiality of components that are known to illustrate medicinal constituents as well as physiological activities [8].

Phytochemical analysis of *Evolvulus alsinoides* (Linn.) Linn.

S. No	Phytochemical compounds	Methanol extract	Ethanolic extract	Aqueous extract
1	Alkaloid	+	+	−
2	Carbohydrate	+	+	−
3	Steroids	−	−	−
4	Saponin	−	+	−
5	Tannin	+	+	+

S. No	Phytochemical compounds	Methanol extract	Ethanolic extract	Aqueous extract
6	Pseudo Tannins	+	–	–
7	Chlorogenic acid	+	–	–
8	Flavones	–	+	–
9	Flavonoid	+	+	–
10	Coumarin	–	–	–
11	Anthocyanin	–	–	–
12	Phenol	+	+	+
13	Terpenoids	+	+	+
14	Triterpenoids	+	+	+
15	Resins	–	–	–
16	Volatile oil	+	+	+
17	Glycosides	–	+	–

+ = *present;*
- = *absent.*

4. Analysis of heavy metals

Test for the presence of heavy metals is done for this plant's leaf extract. Sodium (Na), Potassium (K), Phosphorus (P), Manganese (Mn), Iron (Fe), Calcium (Ca), Zinc (Zn), Copper (Cu), Nickel (Ni) and Magnesium (Mg) were the heavy metals present in the leaf extract. The recent studies have proved that Potassium (approximately 133 ppm) is higher when compared to other heavy metals present [8]. Sodium, Phosphorus, Manganese, Iron, Calcium, Zinc, Copper, Nickel and Magnesium are present in the leaf extract. Metals like Lead and Mercury were not detected in the extract.

5. Antimicrobial activity

Antimicrobial activity attributes the extracts potential to act against the micro-organisms by inhibiting the growth and the activities of them. *Evolvulus alsinoides* (Linn.) Linn. displayed commendable potential of antimicrobial activity. The zone of inhibition of the expansion of the isolates refers to the relative antimicrobial potential that *E.alsinoides* (L.) L. possesses. The results of the antimicrobial activity of *Evolvulus alsinoides* (Linn.) Linn. methanol, ethanol and aqueous extract are explained as follows: [8].

5.1 Antimicrobial activity of *Evolvulus alsinoides* (Linn.) Linn. aqueous extract

The organisms used for the antimicrobial activity of *E.alsinoides* (L.) L. aqueous extract are *Salmonella typhi, Acinetobactor baumannii, Staphyloccus aureus, Klebsiella pneumonia, Escherichia coli* and *Pseudomonas aeruginosa.* The inhibition values for three different concentrations for each microbe were obtained for the aqueous extract of *E.alsinoides* (L.) L. The three different conecentrations are 50 μg/ml, 100 μg/ml and 150 μg/ml. The respective inhibition values (mm) are 8, 11, 14 (for *Salmonella typhi*), 10, 13, 16 (for *Acinetobactor baumannii*), 10, 11, 12 (for *Staphyloccus aureus*), 7,8,9 (for *Klebsiella pneumonia*), 8, 10, 12 (for *Escherichia coli*), and 6,8,9 (for *Pseudomonas aeruginosa*).

5.2 Antimicrobial activity of *Evolvulus alsinoides* (Linn.) Linn. methanol extract

The organisms used for the antimicrobial activity of *E.alsinoides* (L.) L. methanol extract are *Candida albicans, Aspergillus niger, Staphyloccus aureus, Vibrio cholera, Bacillus megaterium, Klebsiella pneumonia, Salmonella typhi, Yersinia enterocolitica, Bacillus subtilis* and *Listeria monocytogenes*. The inhibition values for three different concentrations for each microbe were obtained for the methanol extract of *E.alsinoides* (L.) L. The three different conecentrations are 50 μg/ml, 100 μg/ml and 150 μg/ml. The respective inhibition values (mm) are 12, 15, 21(for *Candida albicans*), 12, 14, 17 (for *Aspergillus niger*), 12, 15, 18 (for *Staphyloccus aureus*), 13, 16, 22 (for *Vibrio cholera*), 10, 13, 16 (for *Bacillus megaterium*), 11, 15, 17 (for *Klebsiella pneumonia*), 14, 17, 18 (for *Salmonella typhi*), 13, 15, 17 (for *Yersinia enterocolitica*), 11, 12, 13 (for *Bacillus subtilis*) and 10, 12, 15 (for *Listeria monocytogenes*).

5.3 Antimicrobial activity of *Evolvulus alsinoides* (Linn.) Linn. ethanol extract

The organisms used for the antimicrobial activity of *E.alsinoides* (L.) L. aqueous extract are *Escherichia coli, Pseudomonas aeruginosa, Candida albicans* and *Staphyloccus aureos.* The inhibition values (mm) for the concentration 10 mg/ml, for *E.coli* was 14 and for *Pseudomonas aeruginosa* is 11. There was no inhibition found in *Candida albicans* and *Staphyloccus aureos.*

6. Antioxidant activity

Antioxidant activity is primarily described as the activity of preventing the propagation level in oxidative chain reactions that acts as a constraint of the oxidation of lipids, proteins, DNA and so on. Free radicals are directly scavenged by primary antioxidants. Unlike primary antioxidants, secondary antioxidants preclude the forming of free radicals indirectly through Fenton's reaction [9].

Fluorescence Recovery After Photobleaching (FRAP) assay can be easily reproduced and it is directly associated with molar concentration of antioxidant present. This can be implicated as the extract of *Evolvulus alsinoides* (Linn.) Linn. that may act as a free radical scavenger. The FRAP (Fluorescence Recovery After Photobleaching) assay portrays the scavenging efficiency of the extract (Ethanol) of *E.alsinoides* (L.) L. at five distinct concentrations. They are 200 $\mu g{\cdot}mL^{-1}$, 400 $\mu g{\cdot}mL^{-1}$, 600 $\mu g{\cdot}mL^{-1}$, 800 $\mu g{\cdot}mL^{-1}$ and 1000 $\mu g{\cdot}mL^{-1}$ and the respective optical densities are 0.907, 1.309, 1.397, 1.561 and 1.825 at 595 nm.

7. GC–MS analysis

The prevailing compounds were, Hexadecanoic acid, methyl ester, Benzoic acid, n-Hexadecanoic acid, D-Allose and Cytidine. The presence of various bioactive compounds justifies for usage of plant extract for various ailments by Ayurveda practitioners [10].

The investigation concluded that the stronger extraction capacity of methanol could have been produced number of active constituents responsible for many biological activities. So that those might be utilized for the development of traditional medicines and further investigation needs to elute novel active compounds

DOI: http://dx.doi.org/10.5772/intechopen.96119

from the medicinal plants which may be created a new way to treat many incurable diseases [10].

Biocompounds of *Evolvulus alsinoides* (Linn.) Linn. leaf extract found by GC MS analysis.

S. No	Name of Compound	Molecular Formula	Molecular weight
1	Tricyclo[2.2.1.0(2,6)]heptane, 1,7,7-trimethyl	$C_{10}H_{16}$	136
2	2,4- Imidazolidinedione, 3-Methyl	$C_4H_6N_2O_2$	114
3	4-Heptanone, 2-Methyl-	$C_8H_{16}O$	128
4	Levoglucosenone	$C_6H_6O_3$	126
5	4H-Pyran-4-one, 2,3-Dihydro-3,5-Dihydroxy-6-	$C_6H_8O_4$	144
6	Benzoic Acid	$C_7H_6O_2$	122
7	Cytidine	$C_9H_{13}N_3O_5$	243
8	4-Methyl-2-Oxopentanenitrile	C_6H_9NO	111
9	2,3-Dihydro-Benzofuran	C_8H_8O	120
10	5-Hydroxymethylfurfural	$C_6H_6O_3$	126
11	2-Methoxy-4-Vinylphenol	$C_9H_{10}O_2$	150
12	Alfa.-copaene	$C_{15}H_{24}$	204
13	Cyclohexene,1-methyl-4- (1methylethenyl)-, (r)	$C_{10}H_{16}$	136
14	Caryophyllene	$C_{15}H_{24}$	204
15	1,1'-Bicycloheptyl	$C_{14}H_{26}$	194
16	1,6-cyclodecadiene, 1-methyl-5-m	$C_{15}H_{24}$	204
17	D-Allose	$C_6H_{12}O_6$	180
18	(−)-5-oxatricyclo[8.2.0.0(4,6)]dodeca	$C_{15}H_{24}O$	220
19	Cyclohexene, 1-Methyl-3-(Formylmethyl)	$C_9H_{14}O$	138
20	1hcyclopropa[A]Naphthalene, 1a	$C_{15}H_{28}O_2$	240
21	Dotriacontane	$C_{32}H_{66}$	450
22	Diethyl phthalate	$C_{12}H_{14}O_4$	222
23	Tetradecanoic acid	$C_{24}H_{48}O_2$	368
24	Tricyclic [2.2.1.0(2.6)]Hept-3-OL	$C_7H_{10}O$	110
25	Benzoic acid,2,6-bis Trimethylsil	$C_{16}H_{30}O_4Si_3$	370
26	2,6,10 trimethyl, 14-ethylene-	$C_{20}H_{38}$	278
27	Cyclohexanol, 4-[(Trimethylsilyl)oxv]-,cis	$C_9H_{20}O_2Si$	188
28	2-cyclohexen-1-Ol,2,4,4- trimethy	$C_{16}H_{24}O_2$	248
29	Pentadecanoic acid	$C_{15}H_{30}O_2$	242

Biological Activity of compounds found in methanolic leaf extrac of *Evolvulus alsinoides* (Linn.) Linn. by GC–MS.

S.No	Compound name	Nature of Compound	Biological activity
1	n-Hexadecanoic acid	Palmitic acid	Antioxidant, Hypocholesterolemic nemaicide, pesticde, Anti-androgenic flavor, hemalytic, 5- Alpha reductase inhibitor
2	Hexadecanoic acid, methyl ester	Fatty acid ester	Antioxidant, Antimicrobial Hypocholesterolemic, Antiandrogenic, Hemolytic, Alpha Reducatase inhibitor.
3	Benzoic acid	Benzen	Arachidonic acid-Inhibitor, Increase Aromatic Amino Acid Decarboxylase Activity and Inhibit Production of Uric Acid
4	D-Allose	Aldohexose sugar	Alcohol-Dehydrogenase-Inhibitor, Anticancer (Duodenum), Antidote (Diazepam), Antidote (Digoxin), Antileukotriene-D4, Circulatory-Depressant, CNS-Depressant and Coronary-Dilator
5	Cytidine	Nucleoside molecule	Glutamatergic antidepressant drug

8. Effect on marker enzymes in Streotozotoc induced Diabetic rats

Diabetes mellitus, generally known as diabetes, is a heterogeneous metabolic disorder indicated by a basic feature of chronic hyperglycaemia with a bother of fat, carbohydrate and protein metabolism. One of the prominent causes of anguish and mortality worldwide is Diabetes mellitus. Pancreatic cells are damaged by creating profuse reactive oxygen species (ROS) that concluded in diabetes mellitus. An imbalance in oxidant and antioxidant activities, Oxidative stress is also included in pathogenesis and complications of diabetes mellitus. The activities including detoxification, storage, metabolism, excretion of the metabolites and xenobiotics performed by liver is vital, thus liver is a crucial organ of the body [11].

The studies show that effect of extract of whole plant of *E.alsinoides* (L.) L. can be availed for the normality of marker enzymes that implies the normality of liver and kidney. The markers AST (aminotransferace), ALT (alanine aminotransferace), ACP (acid phosphatase), LDH (lactate dehydrogenase) and ALP (alkaline phosphatase) were utilized for the study and increase in them in the circulation of blood illustrates damage in liver of diabetic rats, induced by streptozotocin [11]. The diabetes results in increase in the concentration of AST, ALT, ACP, ALP and LDH than in normal conditions. The extract administered was noticed to preserve the values of the enzymes to normal values. Thus showing no crucial difference in between the control (glibenclamide) and the plant extract alone group rats in the normal levels.

The following tables concludes that ethanolic extract of *Evolvulus alsinoides* (Linn.) Linn. importantly decreases the concentration of marker enzymes that exist in serum and tissues, which are commonly found to be higher in levels of streptozotocin induced diabetic rats. Its constructive impact of biomarker enzymes could portray a defensive mechanism contrary to the toxic effect of free radicals consisting in diabetes mellitus [11].

9. Effect of ethanolic extract of *Evolvulus alsinoides* (Linn.) Linn. on marker enzymes as in serum of control and experimental group rats

The results show that the effect of *Evolvulus alsinoides* (Linn.) Linn. is similar and in some cases better than glibenclamide, a drug used for diabetic control. The extract alone also shows notable effects against diabetic rats. The diabetes results in increase in the concentration of AST, ALT, ACP, ALP and LDH than in normal conditions. The extract administered was noticed to preserve the values of the enzymes to normal values. The concentrations of markers AST, ALP, ACP, ALT and LDH (μmoles/L) are 40, 43, 10, 32 and 44 respectively in control. The concentrations of markers AST, ALP, ACP, ALT and LDH (μmoles/L) are 60, 120, 23, 74 and 85 respectively in diabetic control. The concentrations of markers AST, ALP, ACP, ALT and LDH (μmoles/L) are 43, 45, 12, 45 and 45 respectively in Diabetic and *Evolvulus alsinoides* (Linn.) Linn. The concentrations of markers AST, ALP, ACP, ALT and LDH (μmoles/L) are 41, 43, 11, 38 and 44 respectively in Diabetic and glibenclamide. The concentrations of markers AST, ALP, ACP, ALT and LDH (μmoles/L) are 40, 43, 10, 34 and 44 respectively in control.

10. Impact of ethanolic extract of *Evolvulus alsinoides* (Linn.) Linn. on marker enzymes in liver of control and experimental group rats

The research studies conducted portrayed that *Evolvulus alsinoides* (Linn.) Linn. ethanol extract has inhibitory effect not only on alpha glucosidase but also alpha amylase. The inhibitory activity exhibited concentration reliant reduction thus the higher the concentration higher the inhibitory activity. The topmost concentration 100 μg/ml is the concentration that showed the highest inhibition of closely to 63% and the standard acarbose demonstrated the inhibitory activity of 74%.

11. The inhibition (%) of alpha glucosidase by ethanolic extract of *Evolvulus alsinoides* (Linn.) Linn.

In comparison with the inhibition of alpha amylase by *Evolvulus alsinoides* (Linn.) Linn., the inhibitory action of it on alpha glucosidase is limited. The inhibitory activity of alpha glucosidase by the extract of *E.alsinoides* (L.) L. is about 53% at a concentration 100 μg/ml. This study proves that the *Evolvulus alsinoides* (Linn.) Linn. contains therapeutic antibiotic impact in type II diabetes mellitus [12].

12. A cognitive enhancer of spacial memory

D-Galactose caused mental retardation and cognitive dysfunction as measured by open field, avoidance/escape, T-maze, Y-maze and Morris maze in mice. D-gal, a reducing sugar which can be metabolized at normal concentration, induces the production of reactive oxygen species (ROS) and advanced glycation end products (AGEs) at higher doses that causes detoriating effect in the mice.

An extract from *Evolvulus alsinoides* (Linn.) Linn. have illustrated the strongest memory enhancing property by keeping in control of acetylcholinesterase and also increasing the levels of acetylcholine in the brain and so it possess reasonably acceptable antiamnesiac outcome in animal neurotoxicity model. Scopoletin, a furanocoumarin found in *E.alsinoides* (L.) L. promotes the gene and it can be used for monotherapy for learning and memory.

13. HPTLC fingerprinting analysis (high performance thin- layer chromatography)

The TLC chromatogram was run for *E.alsinoides* (L.) L. along with standard for various profiles such as steroids, glycosides and terpenoids.

Glycosides comprise a very wide range of compounds that are of common and ubiquitous occurrence in almost all plants. Many plants store medicinally important chemicals in the form of inactive glycosides. The nonsugar portion contains the biochemically active properties of medical interest. Once the glycoside is split into its two components, the nonsugar component is free to exert its chemical effects on the body. For example, digitalis is a glycoside that causes the heart to contract (pump) more forcefully when ingested [13]. These pharmaceuticallyvaluable glycosides contribute to almost every therapeutic class, cardiac drugs, laxatives, counterirritants, analgesics, renal disinfectants, antirheumatics, anti-inflammatory, antituberculosis, expectorant and antispasmodic action [2].

14. Ethnopharmacological activities

To carry out enzyme inhibition studies, lactate dehydrogenase was cloned from *Plasmodium falciparum* 3D7 strain using expression vector pET28a and expressed in *Escherichia coli* BL21 (DE3).

Methanol extract of *E.alsinoides* was tested at 50 mg/mL concentration for PfLDH inhibitory activity. Methanol extract of *E.alsinoides* (L.) L. reduced PfLDH activity to (25.04 ± 0.51) %. Effects of *E.alsinoides* (L.) L. were statistically significant at 0.001 levels [7]. The chloroform and ethyl acetate extracts were showed graded dose response. The chloroform, ethyl acetate extracts were protected 27.4% and 28.3% at 120 min, it indicates that *Evolvulus alsinoides* (Linn.) Linn. extracts are considerably reduced the inflammation but when compared with standard drug at 12.5 mg/kg body weight was not that much potent drug, but the extracts were reduced the inflammation caused by prostaglandins, histamine and 5- hydroxy tryptamine at initial stage.

15. Conclusion

The traditional methodology utilized variety of medicinal herbs and one of the notable herbs is *Shankhpushpi* also known as *Evolvulus alsinoides* (Linn.) Linn. has been employed in clinical use for centuries. *Shankhapushpi* is contemplated as *Medha Rasayana* attributes to a drug that revivifies, retains and potentiates intellect and cognizance. *Evolvulus alsinoides* (Linn.) Linn. possess activities like antibiotic, antimicrobial, antiinfalamatory and antidiabetic. The heavy metal analysis shows the presence of notable metals in the extract and phytochemical analysis portrays the presence of crucial phytochemicals present in it. Various studies proved that *E.alsinoides* (L.) L. carries nootropic effect and it is indeed a potent plant for the treatment of brain disorders.

DOI: http://dx.doi.org/10.5772/intechopen.96119

Author details

Harshavardhini Kuppusamy R, Jananipriya Srinivasan, Pavithra Sundaramoorthy* and Kannan Kilavan Packiam
Department of Biotechnology, Bannari Amman Institute of Technology, Sathyamangalam, India

*Address all correspondence to: pavithramks@bitsathy.ac.in

References

[1] Dr. V.G. Indhumol, Dr. H.R. Pradeep, Dr. C.K. Sushrutha, Dr. T. Jyothi & Dr. M.M. Shavas, 'Ethnomedicinal, Phytochemical, and Therapeutic applications of Evolvulus alsinoides Linn. – A Review', SOUSHRUTAM An International Research Journal of Pharmacy and Plant science, 2013; Volume 1(2), ISSN: 2320-5830.

[2] Gomathi Rajashyamala L, Elango V, 'Identification of bioactive components and its biological activities of Evolvulus alsinoides linn. -- A GC-MS study', International Journal of Chemical Studies 2015; 3(1): 41-44.

[3] A. NAHATA AND V. K. DIXIT, 'Spectrofluorimetric Estimation of Scopoletin in Evolvulus alsinoides Linn. and Convulvulus pluricaulis Choisy', Indian Journal of Pharmaceutical Sciences, 2008; 834-837.

[4] Frantisek Cervenka, Vit Koleckar, Zuzana Rehakova, Ludek Jahodar, Jiri Kunes, Lubomir Optetal, Randomir Hyspler, Daniel Jun and Kamil Kuca, 'Evaluation of natural substances from Evolvulus alsinoides L. with the purpose of determining their antioxidant potency', Journal of Enzyme Inhibition and Medical Chemistry, 2008; 23(4): 574-578.

[5] K. P. Singh, Bhavna and G. Dhakre, 'Reproductive biology of Evolvulus alsinoides L. (Medicinal Herb)', International Journal of Botany, 2010; 6(3) 304-309, ISSN 1811-9700.

[6] Mukesh Kumar Yadav, Santosh Kumar Singh, J. S Tripathi and Y.B Tripathi, 'ETHNOPHARMACOLOGICAL ACTIVITIES OF TRADITIONAL MEDICINAL PLANT: EVOLVULUS ALSINOIDES', World journal of pharmacy and pharmaceutical sciences, 2016; 5:4 ISSN 2278-4357.

[7] Neeraj Kumar Sethiya, Alok Nahata, Shri Hari Mishra, Vinod Kumar Dixit, 'An update on Shankpushpi, a cognitive boosting Ayurvedic medicine', JCIM, 2009; 7(11) 1001-1022.

[8] Kannan Elangovan, Kumar Supriya, Kandhasamy Murugesan* and Rajamani Aravind,' Screening of Phytochemicals and In vitro Antioxidant activity of Evolvulus alsinoides L.', Journal of Academia and Industrial Research (JAIR),2013; 2:4 ISSN: 2278-5213.

[9] Dejian Huang, Boxin Ou, and Ronald L. Prior, 'The Chemistry behind Antioxidant Capacity Assays', J. Agric. Food Chem. 2005, 53, 6, 1841-1856.

[10] T. K Sujayil and T. S Dhanaraj, 'DETERMINATION OF BIOACTIVE COMPOUNDS IN Evolvulus alsinoides LEAF EXTRACT USING GCMS TECHNIQUE', RJLBPCS, 2016, ISSN 2454-6348.

[11] Duraisamy Gomathi, Manokaran Kalaiselvi, Ganesan Ravikumar, Kanankasabapathi Devak and Chandrasekar Uma, 'Activity of Evolvulus Alsinoides on Marker Enzymes in Streptozotoc on Marker Enzymes in Induced Diabetic Rats', IJPRT,2013; 5:2 119-125.

[12] Duraisamy Gomathi, Manokaran Kalaiselvi and Chandrasekar Uma, 'Invitro α - amylase and α – glucosidase inhibitory effects of ethanolic extract of *Evolvulus alsinoides* (Linn.) Linn., IRJP, 2012; ISSN 2230-8407.

[13] D.GOMATHI, M.KALAISELVI, G.RAVIKUMAR AND C.UMA, 'Evaluation of Enzymatic and Non–Enzymatic Antioxidant Potential of Evolvulus alsinoides (L.) L.', Asian Journal of Pharmaceutical and Clinical Research 2012; Vol 5, Suppl 2, ISSN - 0974-2441